Deep Time, Dark Times

WESTERN SYDNEY
UNIVERSITY

Thinking Out Loud: The Sydney Lectures in Philosophy and Society

These annual lectures aim to be theoretical in nature, but also to engage a general audience on questions about politics and society. The lectures are organized by Western Sydney University, in collaboration with ABC Radio National, the State Library of New South Wales, and Fordham University Press.

BOOK SERIES EDITOR
Dimitris Vardoulakis

LECTURE SERIES EXECUTIVE COMMITTEE
Chair: Dimitris Vardoulakis
Diego Bubbio
Joe Gelonesi
Richard W. Morrison

STATE LIBRARY*
NEW SOUTH WALES

Deep Time, Dark Times

On Being Geologically Human

David Wood

FORDHAM UNIVERSITY PRESS

NEW YORK 2019

Fordham University Press has no responsibility for the persistence or accuracy of URLs for external or third-party Internet
websites referred to in this publication and does not guarantee
that any content on such websites is, or will remain, accurate or
appropriate.

Fordham University Press also publishes its books in a variety of
electronic formats. Some content that appears in print may not
be available in electronic books.

Visit us online at www.fordhampress.com.

Library of Congress Cataloging-in-Publication Data

Names: Wood, David, 1946– author.
Title: Deep time, dark times : on being geologically human /
    David Wood.
Description: New York : Fordham University Press, 2019. |
    Series: Thinking out loud | Includes bibliographical
    references and index.
Identifiers: LCCN 2018020384| ISBN 9780823281367 (cloth) |
    ISBN 9780823281350 (pbk. : alk. paper)
Subjects: LCSH: Humanism. | Human beings. | Philosophical
    anthropology. | Climatic changes.
Classification: LCC B821 .W664 2019 | DDC 128—dc23
LC record available at https://lccn.loc.gov/2018020384

Printed in the United States of America

21 20 19   5 4 3 2 1

First edition

to Helen Tartar (1951–2014), in deepest appreciation for her singular gift

CONTENTS

1.   Herding the Cats of Deep Time   1

2.   Who Do We Think We Are?   26

3.   Cosmic Passions   36

4.   Thinking Geologically after Nietzsche   47

5.   Angst and Attunement   60

6.   The Present Age: A Case Study   73

7.   Posthumanist Responsibility   82

8.   The New Materialism   96

9.   The Unthinkable and the Impossible   107

10. What Is to Be Done? Democracy and Beyond   121

     Acknowledgments   137

     Notes   *139*

     Index   *157*

Deep Time, Dark Times

ONE

Herding the Cats of Deep Time

> In history, a great volume is unrolled for our instruction, drawing the
> materials of future wisdom from the past errors and infirmities of
> mankind.
>
> —EDMUND BURKE, *Reflections on the Revolution in France*

> If you free yourself from the conventional reaction to a quantity like a
> million years, you free yourself a bit from the boundaries of human time.
> And then in a way you do not live at all, but in another way you live
> forever.
>
> —JOHN MCPHEE, *Basin and Range*

Never before has a species possessed both a geological-scale grasp of the history of the earth and a sober understanding of its own likely fate. The question that drives this book is whether we can take on board in an affirmative way what such a deep time consciousness teaches us or what such a terrestrial responsibility would look like. Our situation forces us to confront questions both philosophical and of real practical urgency. We need to rethink who "we" are, what agency means today, how to deal with the passions stirred by our circumstances (resignation, anger, despair), whether our manner of dwelling on Earth is open to change, and ultimately—What is to be done? Our future, that of our species, and that of all the fellow travelers on the planet depend on addressing these questions.

Friedrich Nietzsche once wrote a famous essay addressing the burden that history can pose to the affirmation of life.[1] Although he was talking about human history and human life, the dawning of the Anthropocene invites us to expand the scope of the question from human to geological history, and from human life to life more generally.[2] With specific reference to climate change, this sense of debilitating burden is common; it was the predominant response to Al Gore's film *Inconvenient Truth*, for example. The Anthropocene, and the issues surrounding climate change, are sites of not just resignation but also anxiety, fear, hope, and many other passions. And the political space within which such issues might be addressed is equally marked by passion. I take up here the difficulty of the geological expansion with respect to time and history by working through Nietzsche's essay within the framework of the affective, with sideways glances at Heidegger.

Herding the cats of Deep Time is no easy task. I have sought to align various scientific and pre-philosophical angles on history with the many reflective and methodological remarks that need to be attached to them. I provide many lists and some lapses into naivety. But while I lay out various dimensions of Deep Time, the avowedly philosophical dimension is to be found in what Wittgenstein would call the reminders I am assembling. If not quite what Derrida dubs the undecidable, I try to bring out at least the shape of some of the difficulties we face in thinking about Deep Time. And although it must be thought both forward and backward, I begin by focusing mostly on the past.

The Past

But what is the past? I confess, as I am writing, and with authentic Augustinian head scratching, that I cannot get over that today's breakfast, so recently real, is now over. I start with some everyday ruminations. A few weeks ago, a friend brought me two geodes, orange-brown in color and about the shape and size of cauliflowers, which now sit on my porch. A quick Google search informed me that they are each about 350 million years old and have been together, so to speak, all that time. That's about one-tenth of the age of Earth. There are snapping turtles in my pond whose ancestors evolved about 40 million years ago. At the bottom of my garden one can find arrowheads that date back 10,000 years. There is evidence of farming on the ridges around here some two hundred years ago, and a pile of rocks from the Civil War about 150 years old. A neighbor who was born on my farm some eighty years ago recently died. The old barn, designed for loose hay and now obsolete, was probably built after the Second World War. At the edge of the lake, which I dug ten years ago, dragonflies skim the surface for tiny insects. In an instant they themselves are eaten by swooping barn swallows. All this, so far, with little need for memory.

Time is scaled. The past is disclosed both to a close-up and a telescopic lens. It reaches back from a moment ago to unthinkably distant events. It traverses days, personal events, lifespans, wars, the birth of nations, and other momentous historical events. It encompasses the evolution of our own and other species, indeed the origin of life itself.

I am confident about this story because of the evidence presented to me and my fellow humans, to historians and scientists as well as our various personal and collective memories. This evidence is interpreted according to theories that, as they say, have stood the test of time. But these theories are, at least in principle, revisable (or so we learn from history).

This picture of the past would fill out what we might call a nested retrospective temporality. Such a temporality accommodates everything from the fading memory of the present (such as the beginning of this sentence) to the origin of the universe.

These "events" are not just different points on a long line; they also correspond to different scales of significance, different orders of event. This is important, for although we have not begun to problematize the objectivity of time, we have found it hard to resist acknowledging such scales or orders of dimension. There have been genuinely significant events in the past whose identification does not just reflect our parochial interests—geological events such as volcanic eruptions, meteor impacts, ice ages, the formation of continents, and eager predecessors crawling out of the swamps.

The so-called history of civilization is more problematic. We typically and without difficulty navigate with commonsense distinctions between the past memories of people we know and have known, the culturally commemorated past (books, monuments, artifacts), infrastructural realities (roads, buildings, cities, bridges), the past available to archaeology, geology, and so on. Much of this mapping is at some level just there, taken for granted. But human history, especially when dealing with the legitimating impact of cultural memory, spawns occasions for debate and dissent. Invading armies destroy the cultural treasures of their victims. Nations write slanted histories for political purposes. Turkey denies the Armenian genocide. China deleted

Tiananmen Square (1989) from its books. America pretends Guantánamo is an aberration.

These remarks do not themselves seriously question the objectivity of time. But they do make it clear how any idea that we cannot responsibly move forward without looking back must address questions about the scale, scope, and selectivity of our retrospection.

This account may seem aloof and detached. But while history is nested and complex, in some respects it is there to be discovered. Precisely in the name of truth, adopting a certain neutrality reflects the need to move away from a historiography wedded to the unfolding of meaning and progress in order to embrace a genealogical approach that honors contingency, interruption, and discontinuity and is suspicious of classically canonized events. Foucault acknowledged here a certain positivism on his part—that his patient documentation revealed the slower penumbral emergence of standard practices, frames of reference, ways of discursively organizing the world.[3]

Naturalistic Narratives

Yet, if geological time might be said to be nested, as soon as we consider human history, the nest metaphor breaks down. For we find ourselves dealing with multiple incommensurable temporal series, each marked by their own narrative, not to mention complex interpenetrating relations that break down the silos suggested by "nests." Think of the reverberating echoes of historical trauma down the ages.

The simplest *naturalistic* story would arguably have four phases: the creation of the universe/Big Bang; changes of material states (gases to liquids to solids), the formation of planets, the solar system, the cooling of the earth; the dawn of life and evolution; the appearance of hominids/humans, (opening onto another series: Stone Age, Bronze Age . . . the Industrial Age, and today the Digital Age).

An *energetic* history would center on thermodynamics and the appearance of negative entropy, especially in the form of life processes, which contradict the general drift toward disorder. The question then would be how to locate the specifically human form of life, whether as biocentric inclusion (see Aldo Leopold's land ethic, in which we humans are just plain citizens), or as a step beyond mere life, not least in being able to think *about* life. This step might be thought of in terms of technology or information as ways in which negentropy gets harnessed to life. A cyborg future, or one that detaches itself from life itself, would take this a step further, as in Ray Kurzweil's idea of a singularity. An energetic perspective would be one take on how the human is in important respects colored by and infused with what makes it possible, in ways that cannot be dialectically sublated. Think of the idea of emergent evolution, for which something radically new has come on the scene.[4] This opens up the question of whether we are still stardust and how we are connected to our evolutionary past.

There are other naturalistic histories, those that trace the evolution of Life to "Man," and those that trace the history of the humanoid from the earliest hominid, "Ardi," or *Ardipithecus ramidus*, to *Homo sapiens*.[5] This can raise strange passions. After the famous Scopes "monkey trial" in Tennessee (1925), one man reportedly said that he didn't know what all the fuss was about—being descended from apes was one thing, but it was the earlier link to fish that he couldn't swallow!

The history of the West, sometimes confused with the history of civilization, would include such moments as the Garden of Eden, Greek philosophy, the birth of Jesus, the French Revolution, the Industrial Revolution, the death of God, the Information Age, and the Anthropocene.

None of these accounts need be philosophically sophisticated, though they can be taken up by philosophers such as Kant and Hegel. But the very diversity of accounts is the common property of any culture. This bevy of histories, and there are an indefinite number of them, are our legacy in the West. We pick and choose between histories, and/or critically reflect on them. Even "naïve" history is not one thing; its plurality is both qualitative and quantitative. This is the landscape of the past that Deep Time confronts, drawing along its own agenda.

Human Temporality

What of the role of human temporality? The study of history, of the various levels of past we have alluded to, can be compromised by naivety. This happens when we do not recognize the constitutive role of both our fundamental human temporality and specific shaping interests, each of which recede into the background when we presume we are studying the facts of history. It will be Heidegger's claim, for example, that the very possibility of history as a discipline rests invisibly on our existential historicity. He distinguishes between historicity and historicality/historiography.[6] The latter concern the course of events, and our study of them. While historicity has to do with how Dasein, our manner of being in the world, is historically engaged (or not), how it enacts or performs itself in and as history, which is itself a dimension of our fundamentally temporal existence.

If performativity turns out to be constitutive of historical engagement, how should we best accomplish that engagement? In a strong form, this problematic generates Nietzsche's concern that our scientific understanding of history can be a burden that threatens human vitality, *life*. I take up this sense of burden into a broader discussion of emotion, mood, disposition, and the affective. Is there perhaps an intimate relation between (lived) time and mood?

The Present Age

If the present is where we start in any reflection on history, there is something special about this present. Ours is not the first to witness a radical displacement in human beings' relation to the earth and to history; Copernicus and Darwin did too. Nor is it the first time we have become aware of geological time; Hutton highlighted this. Rather, our moment is marked by the birth pangs of the Anthropocene, the idea that the human impact on the planet has reached geological proportions. It is insufficient to see ourselves only in the light of historical development (civilization, conquest, enlightenment) and evolutionary development from other forms of life. Instead, we must have a broader perspective, a geological perspective, one that prompts the idea of Deep Time. To this should be added an increasingly strong sense of a potentially catastrophic future awaiting us. We are surrounded by visions of dramatic change, crisis, and extinction. We are already witnessing the sixth extinction of nonhumans.

A sense of the apocalyptic has itself a long history. An Assyrian clay tablet (circa 2800 BCE) bears the words "Our earth is degenerate in these latter days.

There are signs that the world is speedily coming to an end."[7] And it has been coming to an end regularly ever since, some today waiting for the Christian Rapture. Is this history of the end of time just another history? Should skeptical academic rationalism triumph? What would it be to say, "But this time, it's real"? That is the moral of the tale of "The Boy Who Cried Wolf"!

If we must, in some sense, start with the present, with now, how can this be characterized? Should we draw our preliminary understanding of the present from our surrounding culture, or approach the very idea of the present in a critical way? As an example of the latter, one would have to understand that the present contains within itself various relations to the past (and future), that there are multiple and contested versions (politically, nationally, ethnically, and so forth), and that philosophical characterizations of "the present age" have been offered by Socrates, Kant, Hegel, Marx, Kierkegaard, Husserl, Heidegger, Marcuse, Arendt, Irigaray, Guattari, and Derrida, to name but a few.[8] This trail of accounts of the present age is itself the product of repeated sophisticated critique, not just unreflective cultural malaise. I argue later for the plausibility of a benign blend of some popular activist accounts, such as Naomi Klein's *This Changes Everything*, with the more hermetic diagnoses of a Heidegger.

Contemporary Rethinking of Time

I am not the sole proprietor of the idea of deep time. Ted Toadvine writes: "Our everyday experience of time has been transformed by the scientific discovery of the geological past and predictions of anthropogenic environmental impacts extending into the far future." He continues "the deep past and deep future . . . are . . . radically heterogeneous ruptures within our everyday temporal experience." He calls this "catastrophic intrusion of deep time within lived temporal experience. Anachronicity."[9]

David Morris writes of "An inaccessibly deep temporality." He describes it as "radical contingency", as a transcendental condition of phenomenology. He alludes to the sense *inherent in things*, connected up with "Merleau-Ponty's *etre sauvage*, 'Raw being' and Schelling's barbarian principle."[10] Glen Mazis also speaks of "The depths of time" with reference to geological time, memory, and selfhood.[11]

Some thinkers, myself included, were released from dogmatic temporal slumbers by Husserl, Heidegger (whose views themselves developed over time),[12] and Derrida. Rethinking time has been at the heart of efforts to rework the scope and limits of phenomenology. For example, John Sallis refers to elemental temporality and polytopical time, including lithic time (a time of the earth).[13] Daniela Vallega-Neu's sense of "disseminating time" also gives voice to the (often importantly embodied) constitutive time of subjectivity and cosmic time, as well as all the other times released when these poles melt. Finally, Joanna Hodge's "looping time," the future perfect time of *Nachtraglichkeit*, draws from Derrida.[14]

There are other efforts relevant to rethinking the traditionally humanistic premises of our understanding of time, from Foucault through Steigler, Deleuze, and the new materialists. I discuss these briefly in Chapter 7. Popular books also testify to this broader trend, such as John McPhee's poetic *Annals of a Former World* (1998) on America's geological past (what he calls "deep history") and Henry Gee's *In Search of Deep Time: Beyond the Fossil Record to a New History of Life* (2000).[15] My point here is simply that our time, a time "out of joint," itself harbors many layers of strange loops. These include both the advent of the Anthropocene, as a cultural and scientific arrivant, and this expanding space of philosophical reflection on time. And such reflection is not just "on" time, but adverbially, performatively temporal, and historical, as we will see with Nietzsche and Heidegger.

Heidegger is not just one name among others on this list. After Heidegger, the whole of the epoch of Western history, the reign of "forgetfulness of Being" and *Machenschaft*,[16] is to be understood in terms of another "deeper sense of history, one that would not be chrono-logical, but onto-logical."[17] This deeper sense, in my view, coincides with the sense of man's dominion over nature expressed in Genesis. The Anthropocene would mark materially, ontically, what had long been playing out ontologically. Heidegger's discussion of the Other Beginning projects, imagines, solicits, and performs an alternative possibility[18]. While one could set everything within the drumbeat of the strife of earth and world, I have tried to strike out in ways not wholly dependent on Heidegger's ways of framing things, even if in the long run these ruminations are but one long detour on the path of responding to Heidegger.[19]

Learning from the Past

So how do we learn from the past? History has bequeathed to us fundamentally two different judgments about what we can learn from history, where history usually means human history. The first is ironic pessimism, such as Marx's "History repeats itself, first as tragedy, second as farce" or Shaw's "We learn from history that we learn nothing from history." Hegel, Santayana, and Churchill said much the same. Then there are those for whom history teaches us useful lessons, such as Burke, Hume, Machiavelli, and Kant. Thus, Burke: "In history, a great volume is unrolled for our instruction, drawing the materials of future wisdom from the past errors and infirmities of mankind."[20]

For Nietzsche (at least in *Uses*), it is human history, perhaps even German history, that primarily concerns him.[21] That is to say, events for which 'we' have been responsible, or have been engaged in. The burden would be the dispiriting recognition of what we (we Germans, we Europeans, and by transposition, we British, we Americans, we humans) have been capable of or incapable of. The challenge would not just be to learn from our mistakes, but at the same time to rekindle a sense of joyful agency, unencumbered by the blood on our hands.

In our time, this narrower focus on national history has been evident in postwar Germany, in the aftermath of the Holocaust, and in the ongoing legacy of slavery and the civil rights movement in the United States. These specific histories can themselves then get caught up in wider considerations (and histories) of European anti-Semitism on the one hand and racism on the other. And the question of what we can learn gets closely tied up with questions about what we can remember or allow ourselves to remember. Nietzsche's talk of "active forgetting" suggests a grasp of the traumatic dimension of much of history, which would offer other sobering explanations for the repetition of folly.

The very idea of learning from the past raises a number of questions:

First, how might we depart from the idea of history as an objective record? To what end? What might "learning from history" mean? Are "we" spectators, agents, participants in this process, or what? Who is this "we"? (This word is used throughout this series of questions, largely unscrutinized.)

Second, what would a really critical eye tell us about the "objective record"? For example, what tacit valuations are folded into our thinking about past *extinctions*? Are they such a bad thing? Can we really proceed with such a "precritical" view of time? (Surely we need to recognize the human impact on this very ability to construct a story of the past, to conceptualize it as some sort of continuity, causal or otherwise, and indeed to attribute a continuing "reality" to it. This goes way beyond the specific distortions of particular personal, national, ideological, or cultural takes on the past.)

Third, while it *seems* fairly straightforward to learn from our mistakes (though it is not!), what happens when we expand into evolutionary and geological history? Do we then leave behind the space of our "mistakes" and actions we could replay more wisely? In what sense is the last Ice Age part of "our" history? That it cleared the way for mammals like us? That it reminds us of sweeping climatic change? Should such phenomena open up to a sense of a necessarily *shared* planetary agency?

Fourth, can we square this nested objective past time, in which it seems that the intensity of lived time has been converted into something "dead," with the ongoing vital flux? Should we not let the dead bury the dead? Isn't the past just water under the bridge—"history"? Beyond Nietzsche's worry, in what sense is the past even real?

Fifth, we are said to have entered the Anthropocene, following the Holocene, the age in which man's impact on the planet has taken on geological significance, impacting its fundamental physical processes. How does this affect "our" understanding of (our) history? Have we (re)acquired a species agency? Does this call for a species responsibility, or what Hans Jonas called "ecological responsibility"? In a reprise of the language of a Hegelian Sartre, can we move from being a species-in-itself to a species-for-itself, and then to a species-for-others?

Nests of Time

The past may in general be characterized by sequentiality, but it also gathers itself into nests, shelters. Just as there are places, locales, which shelter relations relatively independent of those that obtain "externally," the same is true of time-shelters.[22] The things on my writing desk form a kind of community quite separate from those on the table on the other side of the room. Analogously, the events in the ninety-plus minutes of a soccer match (passes, goals, fouls, substitutions, saves, protests) have internal relations of meaning and effect initially confined to that period of time. They may spill over. Statistical curiosities are constantly being created—"the first time Grantham have won with a disputed penalty in extra time." The final score may bring league victory, relegation, a fine for misbehavior, or a manager's dismissal. The match is "nested" within other frames of relevance, which presuppose the original time-shelter of the match. If defeat in one match results in relegation, the team will play a new set of teams in a lower league in the following season. And soccer itself could be said to direct and domesticate tribal loyalties with a long history. This example of nestedness is a cultural artifact. But the same kinds of circles of relevance can be found in "nature." Lift a rock in a field and you will typically find a bunch of small creatures living together. Under the next rock, a different grouping, with species overlap. Some may venture out at night and become food for small mammals, entering an adjacent ecosystem. Life rhythms established in the summer may change with the season, demonstrating long-term dependency on the time-shelter we call the earth's movement around the sun.

How does this model apply to Deep Time? Nestedness is meant to capture the idea that one time-shelter can be set within another. And even when there is no obvious outer "shelter" there will be forces and influences from which the shelter itself is relatively immune. But porosity is always there! New Orleans was a regular little nest, with its habits and rhythms, until Katrina hit.

These examples exhibit something of a spatial drift, I admit. Perhaps it is because place boundaries are more tangible than temporal ones. And in truth they may often be conflated. A bustling city is both a place and a polyphonic entanglement of times. The very distinction between space and time is so often artificial.[23]

The implication for Deep Time is this: Our dwelling consists of a collection of nested time-spaces of relevance in which we typically live, move, and have our being. The morning meeting, the day at work, the week, mealtimes, the weekend, the semester, the year, shopping, birthdays, rituals, holidays, and so on. All of these are built up on metabolic rhythms of eating and sleeping, themselves attuned to circadian cycles, to the moon, to the seasons. We could call this ongoing dwelling: business as usual. *Mutatis mu-tandis*, it has always been thus. Even describing these shared spaces of social intimacy, we begin to adumbrate wider temporal horizons when we reference birthdays, rituals, and the seasons. We are caught up in cycles of repetition in which the past lives on. And these very patterns presuppose relative peace and prosperity, made possible by past wars, mass production of goods, long-standing cultural institutions (like universities, often founded on the questionable wealth of nineteenth-century businessmen), energy from fossil fuels, and so on. Our present floats on a many-layered past, with roots sunk deep into it.

Even if we remain committed to a certain phenomenology we need to acknowledge that dwelling has a back door, it is historically shaped, and not just by human history. Meaning is grounded in historical forces, in matter.

The Matter of History

How can we best capture this relation to the past? I began by mentioning many layers of "history" sedimented in the place I live. Many of these layers have left physical evidence on the ground—geodes, a rock pile, traces of ancient field boundaries, live turtles, a crumbling barn. It is easy to confuse the evidence with the history, in two different ways. The geodes are 350 million years old. That means that these very objects existed pretty much as they are now, that long ago—before there was any prospect of this now, or human beings, before the dinosaurs, or their extinction. In other words, this is an objective reference to the past, not just a claim about how old some present thing might be. And yet it is a reference to a past relevant to this present. It explains how this geode comes to be here. It was once "here," in some sense of here, created by volcanic explosion, and it stuck around. The explosion is a discrete past event causally responsible for a present phenomenon. The barn raising of the 1950s was similarly an event that produced the barn we see today. Neighbors gathered, drank moonshine, hauled and hammered, and the barn went up. They did so in expectation of a future—hay that would grow season after season and needed to be stored dry to feed cattle over the winter, and an ongoing community of cooperative relations between neighbors. This past was a real living event with its own temporality, just like this one. They used trees they had cut into boards that had been growing for years in mixed deciduous hardwood forest, drawing on skills passed down for generations. These evident connections are real and important. They give us a taste of what we take to be true—that the past is entirely responsible for the present, and that most of it has left no obvious trace, no continuing object, no signature.

A Certain Realism

Let me confess the flavor of my realism here, without addressing all the metaphysical objections that it surely raises. Stone things are notoriously long lasting. So, just as we can easily confuse legacy objects with the past that bore them, we can easily forget that most of the past is lost, shrouded in the fog of time. And yet no less real. The torture inflicted by the rack in Warwick Castle in the seventeenth century may have gone away, but when it occurred it was agony.[24] That fact has not changed. And while everything around today is built on the past, not only are there pasts that have bequeathed no recognizable present (though they may have played their part) there are other pasts that have made no significant impact at all. The proverbial butterfly on the other side of the world might have made all the difference, but equally its fluttering could have been swallowed up by a great wind. Moreover, if we take seriously the way the past supplies the material conditions of possibility for all that comes after it, the very expression "the past" easily lends itself to misconstrual. If we concern ourselves with the pasts of different places, based on some sense of relevant causal antecedents, albeit with a porosity to wider contexts, then we will need to speak of multiple pasts, not just The Past. Even if we stick to one place, its distinctive past will continually be changing, as every day, every moment, transforms the possible into the actual, and then into what once was. These pasts do not just grow, they may get transformed, as tipping points are passed, the last of a line dies out, and so on. Here we are tacitly giving in to a certain simplifying domestication of the past. But we can go the other direction too, by recognizing the infinite fractal complexity of the real at any one moment— that every past moment could be said to project a multiplicity of adumbrations (both causal possibilities and intentional prospects). For any past that we could conceivably recollect is infinitely surpassed by the "reality" of what happened (including, as it does, what easily *might* have happened, or what was actually imagined or feared).

If Deep Time is tied up with a relation to the past, and if the past consists of previous presents (or ongoing processes "taking place" at particular times) as fractally and horizonally complex as this one, then our relation to the past will depend a great deal on our receptivity to the complexity of our own time. This suggests a countermovement to the more obvious thought that a simplistic understanding of the past may give one unrealistic expectations about contemporary possibilities. May '68 in Paris might have failed as a revolution because people overlooked the perfect-storm convergence of factors that made 1789 possible and were missing in 1968. Retrospective romantic idealization can be blinding.

So much, then, for preliminaries. I began from Nietzsche's worries about the dispiriting burden of history. The evidence of repeated human folly that it contains can thoroughly dampen our sense that the future might be different. And it more generally encumbers our naïve vitality. The poetic body bags of history sap courage, as Plato complained of Homer. But what if the words "history" and "life" in the title were given expanded scope? Deep time instead of history; life instead of human life? Does that just make things worse?

A common reaction to climate change news is resignation. The glaciers are melting—but what can I do? The environmental movement has often been castigated for its pessimism.[25]

The issues raised by climate change are politically unpopular, and hardly figure in our elections. Many are openly hostile even to discussing the issues. In Chapter 4, I try to set Nietzsche's worries about history and life into a broader discussion of the significance of the passions—emotion, mood, dispositions, and time—in the light of Hume's famous claim that "Reason is and ought only be the slave of the passions."

The Ages of the World

At least initially I see a posthumanist history of the cosmos as nested within other histories. Geologically the Anthropocene Age—from the eighteenth century onward? The start date is disputed—follows the Holocene, which follows the Pleistocene.[26] I was tempted, before discovering the massive scope of my own ignorance, to summarize the geological record and then draw out of it what might be relevant to our purpose. Instead, I must report that that task is beyond herculean. Here are some scraps from the failure of that expedition. First, cosmic deep time is made up of different levels of deep time. These include ages, epochs, periods, eras, eons, and supereons, each larger than the next. These cover the creation of the universe (13.8 billion years), the solar system (5B), the earth, from leftovers after the creation of the sun (4.54B), and then the formation of the moon. Life began from .5B to 1.5B later. Early hominids appeared about 2 million, or 2M, years ago, while *Homo sapiens* emerged some 250,000 years ago. The Holocene saw the development of modern civilization starting some 11,700 years ago. And here we are. In lieu of anything systematic I make the following observations:

1. What are picked out by these periods and intervals are massive changes in the shapes, constitution, and distribution of matter. Explosions, gravitational and magnetic forces, the creation of stars and planets, cooling and solidification, shifts in tectonic plates, ice ages, and so on. These mark the creation and transformation of bodies, processes, atmospheres, systems exhibiting stabilities, periodicities, and instabilities, and they supply background conditions for other developments (such as weather and life).

2. We are practically forced into accepting—at some level—a process ontology, in which things are temporary congealings, and, moreover, a version that allows for transformative events.

3. Leaving aside cognitively impenetrable dark matter and dark energy, the numbers (forces, times and distances) of deep time events are beyond our ordinary comprehension. The primal fire that followed the big bang was a billion trillion times hotter than the sun, and a trillion trillion trillion trillion times denser than rock. This is the very same universe in which we sit down for breakfast and drink coffee, and a good example of Toadvine's anachronicity, the interruption of lived time by deep time. If we cannot harness imagination to give life to these mind-boggling numbers, there is perhaps something in Kant's sense that the experience of the sublime has a recoil that, even as its breaks its teeth on the real,

teaches us something about ourselves. Did not philosophy itself begin with wonder?

4. If the big numbers weren't enough, the idea that the cosmos and indeed time itself "began" with an explosive event, that such an event inaugurated all that there is, a truly unthinkable thought, raises the question of whether such unthinkability can be confined to that first moment, or whether in principle it pervades the whole. Jonathan Edwards, for example, claimed that the cosmos is being recreated at every instant.[27] We are in the territory of Kant's first antinomy, with no compass.[28]

5. Nestled in all this activity, the appearance of life seems like an event of a different order.[29] Even if, as Clausius wrote in 1865, "The entropy of the universe tends to a maximum," the emergence of living beings, centers of negentropy, seems to provide local exceptions. They do not defy the second law of thermodynamics, because they are not closed systems (organisms eat and breathe). But the emergence of beings that strive to maintain their negentropic status seems to take to another level the local "order" found in (say) the solar system, in ecosystems, perhaps even in the atom.

6. Even if life and then human life are merely recent blips on the radar, it does not seem like anthropocentrism or covert theology to ask whether something distinctive or "new" is emerging through cosmic evolution. Nietzsche insists that it would be very shortsighted to think of humanity, certainly as now constituted, as the end of the road, let alone the "purpose" of the earth. But if geological thinking seems attuned to a pervasive naturalism, it raises the question of whether that very capacity for thinking could conceivably be contained within such a naturalism, or whether consciousness or self-consciousness is, as it were, another big bang, another beginning.

7. One can pass through the education system with little understanding of deep time. And yet its story is readily accessible. Consider the mass-market book *A Walk through Time: From Stardust to Us*, subsubtitled *The Evolution of Life on Earth*. It is a celebration of terrestrial creativity in deep time, firmly avoiding teleology or covert theology, crammed with detail. It is not philosophy, but reading it filled a hole in my being that I didn't know was there, a supplement to what had seemed already complete.[30] It was a timely reminder of the importance of scientific knowledge and

detailed narrative for philosophical reflection, even if we need to subject to further scrutiny the frames of reference in which they are presented.

A Temporal Phronesis

Thinking about time at all is as difficult as Augustine said it was. Thinking about Deep Time compounds the problem. Geological time necessarily folds together our human grasp of (or fumbling with) a vast past and a frightening future, on the one hand, with the reality and efficacy of that past, whether or not we can grasp it at all, on the other. A sublime drama. As my last methodological foray here I suggest that, concerned as we are with deep geological time, a critical consciousness needs to absorb a certain temporal phronesis, or facility with thinking about time in all its multistranded complexity.

The project it serves is nicely captured by Tamsin Lorraine (inspired by Deleuze), when she writes of "attuning ourselves to the multiple durations of our lives in ways that allow us to skillfully unfold the creative possibilities of the multiple assemblages of which we form a part rather than fixate on our representations of life."[31] I would tweak this by adding that the "multiple durations of our lives" are now stretched out against the sky, extended, exploded, onto a geological frame. And that the meaning of "our" and "life" is fundamentally in question. In thinking about these multiple durations, in thinking geologically about "our situation," riding the climate-change tiger of the Anthropocene, it is worth articulating some of the concrete shapes of time with which we need to be able to work.

1. Irreversibility. There are interesting ways in which one can change the past—for example by marrying the person you met on the plane—which turns that meeting into "the first time we met" by constituting a continuing we. And we often take for granted that events are reversible. We mend things we break, we apologize for bad behavior, we repay loans. But climate change is for all intents and purposes irreversible. And species loss too; extinction is forever. It is not clear that we have grasped this.

2. Fatal delay. Sometimes time is of the essence, as they say. If the boat has left the harbor, you missed it. Preventing catastrophic climate change means acting before we have absolute proof that we need to. It would not

be rational to wait for the tumor before stopping smoking, and yet this is the shape of our response to global warming. And delay does not just happen on the human side. Many of the changes to the atmosphere are already in the pipeline, programmed in as delayed effects.

3. Structural inertia. We find ourselves saddled with a toxic economic system that would have to change before the change we need can happen. It is hard to see how to change the direction of this tanker.

4. Uneven development. There is a mismatch between our current technological capacities and the social (and political) institutions needed to manage the consequences of deploying them.

5. Tipping points. We act as if change was linear, when it is often best described by catastrophe theory. At a tipping point, dramatic nonlinear events occur. We go to war, we snap, the water overflows the glass. Tipping points are often examples of a broader shape that we could call the open future—a recognition that induction may no longer help us, that there is genuine uncertainty about what may happen. (See Derrida on 9/11 and modern angst, captured in Lars von Trier's film *Melancholia*.)

6. Time unimaginable. I began this introduction with a miscellany of levels of pastness; here is a second loop of the spiral. I used to live in sight of Warwick Castle, built in 1068. I can just about think back to the American Civil War, which ended in 1865. The age of my farm is about a hundred years. My puppy is about four months old. All these are roughly imaginable times. But when I am told the age of the universe is 4.5 billion years, something gives. I thought I had a handle on it when I was told that the coral fossil still traced on my limestone rocks is some 450 million years old. I thought—well the universe is just ten of those; I can count them on my hands. But could I honestly tune in to the original 450 million? No I could not. It's at this point that I have some sympathy with Archbishop Ussher, who proclaimed the universe to have begun in 4004 BCE, with God carefully depositing fossils for our amusement. His time-line seems manageable. But as much as I cannot imaginatively grasp billions of years (even indirectly), I can, with a touch of the sublime, recognize that I cannot grasp it. At the other end of the spectrum we might refer again to messianic time from our previous list—anticipating the impossible, the unimaginable, an event for which we have no concept.

7. Multiple strands. We are participants in and witnesses to multiple processes, each of which have their own time horizons and which spawn their own narratives (some overlapping, others seemingly independent). There is "my day," "my life," "my historical consciousness," not to mention "the gentrification of our neighborhood," "the decay of the public sector," "the evolution of mammals," "the birth of galaxies," and so on. Scale and material qualities separate and entangle these different times.

8. From start to finish. We are now in a position to grasp, from a geobiological perspective, both the story of our origin as humans and quite possibly of our own demise (if not as a biological species, at least as a dream, an imaginary, a promise). This temporal framing of the human project is (almost) unprecedented. Remember Foucault: "man [will] be erased, like a face drawn in sand at the edge of the sea".[32]

9. Aporetic time. The idea that time might be aporetic—that we might find ourselves existentially and/or reflectively unable to piece together our temporal dwelling into a coherent picture—might be thought to be an acceptance of failure. Instead, I suggest that we can anticipate, and even welcome, this. Trauma, for example, profoundly disturbs a certain normalized temporality, and yet it is common. That time is out of joint challenges only our desire for a certain reductive simplification of the temporal.

These are each little shapes, molecules, templates, apps of time. Even when they cannot be detected ahead of time, we arm ourselves better when we can recognize their likely or possible emergence. I am claiming here that dealing with, or responding to, geological time, requires in part the articulation and development of a range of time heuristics, both general and specific. That time, at least in this sense, should turn out to be multiple is something Nietzsche ought to embrace.[33]

Science and Climate Science

So, what can we know about deep time? Much of our purported *knowledge* about the past is the product of various sciences, and yet each such science rests on fundamental ideas that undergo paradigm shifts over time. Should we believe the new kid on the block, climate science, especially as it attempts to predict an unprecedented future, where there can be no real comparables? The worry is magnified if we believe Nietzsche that truth is essentially perspectival and Heidegger that science does not think.

I claim that we can be realists in science, even as we embrace various kinds of critique of the very enterprise, whether phenomenological, genealogical, or deconstructive. Science may not think, but we do not need aerodynamics to think for our plane to take off from the runway. Birds in the turbine intake do not discredit the science; they mess up the engine. As philosophers with phenomenological or postphenomenological sympathies, we have to remember that there is a difference between Archbishop Ussher (for whom the earth was recently created, an empirical claim) and Husserl (claiming "the earth does not move", a claim beyond the empirical). Husserl's claim is in a different register from Ussher's. Whatever his motivation, Ussher marshaled empirical evidence and was wrong. Climate science can err too. Unanticipated feedback loops might cancel out predictions based on linear developments. But climate science is perfectly aware of nonlinear change—catastrophic tipping points such as the dreaded 2 percent rise in the average surface temperature of Earth. And the basic science of climate change is uncontested—that CO_2 is a greenhouse gas, greenhouse gases trap solar radiation, and increases in the surface temperature of Earth could dramatically destabilize weather patterns—these are not contested. If there is doubt, it would be about the timing of these changes.

The question, in my view, is not "What can I know?" but "What can *we* know?" And then: "How are 'we's' constituted?" and "What does it means to know something and not act on it?" These questions are addressed in Chapters 6 and 9.

After Humanism?

Does Deep Time enable us to think through the various thorny questions tied up in anthropocentrism or humanism? Or, conversely, do we have to sort them out first to adequately address Deep Time? Nietzsche takes us part of the way. To get anywhere, we must jettison any teleological sense of the human and any sense that deep time is a long preparation for the advent of man. Man is a rope across an abyss, as he puts it.[34] In their different ways, Heidegger and Deleuze take this further, while Derrida grapples with Heidegger's messy attempts to step back from humanism. Dasein as "Openness to Being" is arguably both a displacement of the human and its reinstallation. Might it not be that such a double inscription is unavoidable, that the most strident antihumanism would embody a thinking that performatively outstripped its own thesis? I am persuaded by Heidegger's sense that serious thinking demands that we have skin in the game, that we bring to the table a real engagement with the question of Being.[35] We do not stop being human when we dissociate ourselves from deeply entrenched projective illusions.

There is no need for hysteria when stepping back from our metaphysical legacy. We do not need, for example, to refuse all reference to experience, not least because experience can be transformative of the subject that undergoes it, as Hegel knew. It can also be a vehicle of subversion, as when Blanchot speaks of the experience of the impossible, as well as the impossibility of experience. When Derrida refuses Nancy's offer of a "provisional" Subject, he is making a strategic choice, not an ex cathedra pronouncement.[36] Kristeva's poetic displacements of the subject position, Heidegger's musings on the middle voice, new materialists' extension of agency to matter—all proceed on the same path. Concepts, narratives, and traditional distinctions have powers that can be redeployed. We can put new wine into old bottles. Everything hangs on strategy, on how this is done. Adverbial posthumanism! And yet success is never guaranteed.

I do not think we can resolve the question of priority—whether we need to exorcise our residual humanism to think about Deep Time, or vice versa. Our wager here is that a critically reflective interrogation of Deep Time will provide more ammunition for some sort of posthumanism.

Does Heidegger Help?

Consider Heidegger's remarks on the sun.[37] These passages shed fascinating light on Heidegger's attempt to ground natural time on Dasein's existentiality. For although everyday Dasein reckons with time, clocks, and timetables, this reckoning is not originally calculative, but is tied up with

Dasein's being as care, as thrownness, as entangled existence. Whatever derived forms of time measurement we come up with—like measuring the length of our own shadow or consulting the watch on our wrist—they all stem from the rising and setting of the sun, from which comes the "day" in our everydayness. And because this horizon is one we share with others living under the same bright sky, this orientation to and by the sun grounds our being with others. Exchanges about the weather foster much phatic communion, especially in England.

One could argue that in this passage another possible Heidegger can be glimpsed, or perhaps the unthought of Heidegger himself. (It was he who wrote: "The unthought is the greatest gift a thinker can bestow."[38]) For, as Plato, Hegel, Nietzsche all knew, the sun is both an opening onto much of what philosophy holds dear, and perhaps the site of a certain refusal. The sun, at one level, is but another thing in the sky, and yet it is the deep ground of visibility itself, or our even having eyes to see. It is the source of the energy that fuels the earth's processes, life, and then consciousness indirectly. The Sun is the ultimate hybrid being—material yet transcendental, supplying conditions of possibility for much that matters. And when one considers that our climate crisis has its own basis in the burning of fossil fuels (deep time reservoirs, standing reserves of ancient solar energy), and that global warming is the product of the greenhouse effect (the rates of absorption of daily solar radiation), it is clear that the Sun is the return address of much philosophical mail. If philosophy itself relies on maintaining something like the ontological difference, or a gap between material and transcendental, arguably the *event* of the Anthropocene is that it blows this rigid distinction out of the water.

Heidegger wants to show that historical thinking (historiography) presupposes our inherent historicity (our being-historical). But might not that thinking itself be dependent on our material history in ways not immediately available to us? The question more generally addresses the logic of dependency—what sort of claim material conditions of possibility can (or should) make on what they make possible, and how that question is to be answered. Derrida dramatizes some of these same issues in his discussion of the animal, especially *The Animal That I Am*, where the following (*Suit*) marks a dependency on the animal (evolutionary?) unacknowledged in the mere "I am" (*Suis*).[39]

The question, put abruptly, is this: Does heteronomy undermine autonomy or make it possible? Here Heidegger's essay "Language" offers a subtle way forward. It begins with: "Man speaks" (quoting Humboldt). Heidegger takes speaking in a broad sense at the beginning, but by the end, the sentence gets reversed: "Language speaks." And then he adds, "Man speaks when he listens to the voice of Language."[40] This is not just Saussure's sense of the dependency of *parole* on *langue*. Rather it means that autonomy is made possible by receptive engagement with conditions of possibility, rather than their denial or refusal. That is the bet I am making with Deep Time. If it generates a labyrinth of undecidability, it is one we need to go through, but never quite have done with. Ethically, we are at our most human not when we claim all our supposed entitlements and privileges but when we are open to various forms of dispossession; the possibility of that openness may be distinctive. It might be said that this is not real autonomy. But what could be more fictional than an autonomy that presupposes background conditions it cannot acknowledge!

Another Displacement of Man?

Surely Deep Time is but one of a number of displacements of Man (or a certain understanding of Man) as the center of things, supplying yet another history for a list that includes Copernicus, Darwin, Marx, and Freud. Each time there is resistance—cognitive, affective, legal, political, as well as philosophical. Arguably the microbiome project continues this chain of displacements. We have only just digested our ape ancestry, and now we are told we are mostly bacteria. We can now ask a Derridean question about hospitality regarding Deep Time. What would it be to welcome this? Must I start saying we? What would it be to welcome being made of bugs or stardust? What does welcome mean anyway? We do have other options—think of Nietzsche's account of how we could respond to the Eternal Return: "Would you throw yourself down and gnash your teeth and curse that demon? Or would you answer, 'You are a god. Never have I heard anything more divine'?"[41] The latter is a welcoming in which the welcomer rises to the occasion.

    Resistance to history is common, if not simple. Holocaust deniers clearly have major investments in their position, as do creationists with respect to evolution. These are easy cases. But what about the denial on the part of the United States of slavery or Native American genocide? Or its current lemming style over-the-cliff leadership in global warming? Or its responsibility for the rise of international terror? Its promotion of a lethal neoliberal global trade agenda? The United States? What is that?

Welcoming Deep Time

If hospitality is the converse of resistance, what would it look like in the case of Deep Time, and specifically the Anthropocene? If my gesture toward opening up Heidegger's reference to the sun is any guide, hospitality to Deep Time would involve redrawing the boundary between material and transcendental, or showing perhaps that it always operates locally. (Think of the Moebius strip—locally two sides, but forming a continuous surface.) More broadly, we might conclude that it would involve giving a similar treatment to embodiment, animality, life, and agency. Each time (witness Heidegger) we are tempted to a reversal, an exclusion, a reinscription of an opposition. The point of deconstruction was always to open up such scenes to other possibilities, to welcome the Other in various ways. It is unclear that musing over pure hospitality is ever especially productive. More fruitful, for example, is the recognition that when dealing with traumatic shock (such as these various displacements of Man), neither mourning nor melancholia has all the answers. The first would work them through without remainder, the second would never let us forget. This issue will return in thinking through the Nietzsche essay.

These have been a series of remarks, considerations, cautions, and dimensions to bear in mind when thinking about Deep Time. If we follow these recommendations, we will address the question more adequately. But Deep Time and the dawn of the Anthropocene stir deep feelings: angst, incredulity, humility, denial, shame, defiance, resignation, and more.

Every age raises deep philosophical questions in its own way. For Luce Irigaray, sexual difference was the question of the age. And war, freedom, justice have all had their day. None of these questions have gone away, but today the spectacle of anthropogenic climate change is pressing philosophy to the limit. Agency, responsibility, time, history, nature, life, science, even

truth—are not only live issues, they are becoming perspicuously mortal concerns. So too is how we deal with the passions aroused by our situation,
which both drive and block an adequate response to it. Contra Hume, reason cannot *just* be "the slave of the passions." To show why, I follow up Nietzsche's sense of the oppressive burden of the past in the light of our legacy of
cosmic and other philosophical passions, the particular challenges of "the
present age," and the broader relation between time and affectivity. But
I will first address the constituency question: Who do we think we are? Are
we just one species among many? Could we embrace the rest of life on the
planet in a collective *we*? Do *we* philosophers too easily slip into speaking
for others? Can new *we*'s (of shared agency, resistance and responsibility) be
constituted?

Who Do We Think We Are?

> Man is a rope stretched between the animal and the Superman—a rope over an abyss.
>
> —FRIEDRICH NIETZSCHE, *Thus Spake Zarathustra*

> "Man" [could well be] erased, like a face drawn in sand at the edge of the sea
>
> —MICHEL FOUCAULT, *The Order of Things*

So, who do we think we are? The last decades have witnessed a gathering skepticism about traditional humanism. The fundamental objection is that the very idea of Man often serves as ideological cover for domination by race, gender, culture, or species. "Man" is a loaded term, caught up in various legitimation narratives. Yet we are a distinct species: *Homo sapiens*. While acknowledging the varying impacts of different farming methods, urban life, indigenous lifestyles, hunting practices, and so on, we humans are responsible for a sharp rise in the extinction of other species on the planet. The animal man has a biological reality beyond color and creed. What, then, can be said about the distinctive features of human beings as a species? How far can we get in giving such an account without introducing normative considerations? And as human beings ourselves, can we avoid covertly blowing our own trumpet?

Consider first our inheritance. *Homo sapiens* is an essentially historical creature, both in ways shared by other living beings and in other more distinctive ways. *Homo sapiens* survives as one of a group of hominids, the others existing now only as fossils or genetic legacy. After we split from the apes tens of millions of years ago, some twenty types of hominid followed: *Australopithecus afarensis, Homo habilis, Homo erectus*—the list goes on. *Homo sapiens sapiens* (modern humans) itself overlapped (and interbred with) *Homo sapiens neanderthalensis* (Neanderthal humans). As hominids succeeded one another, typically brains got bigger (but not always) and bodies got weaker. We evolved toolmaking skills, then moved from hunting and gathering to farming, and now there is a seemingly inexorable drift toward urban life. Our diets, beliefs, and social arrangements followed suit. We are not unique as toolmakers, as birds prove when they use twigs to extract insects from trees. But our development of technology has taken toolmaking to a wholly new level, dramatically affecting communication, transportation, culture, agriculture, the production of material goods, and lethal weaponry. Meanwhile, we have developed the most sophisticated and creative symbolic universe, through writing, art, music, theater, film, and so on. A "second nature," it almost rivals the natural world in its wealth and complexity. We may be said to have realized what Teilhard de Chardin called the "noosphere."[1] However, as Freud noted in a 1932 letter to Einstein cultural restraint has lagged behind advances in the technology of aggression. As a species, we are like kids with AK-47s. We set up institutions (like the UN) to bridge the gaps left by uneven development, but international cooperation is itself a tender shoot facing heavy frosts.

We do not merely have an evolutionary history, we have oral and written histories, which offer opportunities for narrative orientation: progress, decline, salutary lessons, dramatic reassessments and inspiring examples. In an increasingly globalized world, Western grand narratives of unification are fraying. It is harder than ever to treat Europe as taking up the Greek ideal of Man, and carrying its flaming torch forward, or to deem America its worthy successor. Triumphalist claims about the end of history (Fukuyama) look increasingly fatuous. In this light, it is not hard to conclude, in yet another narrative, that *Homo sapiens sapiens* is not actually smart enough to override more deeply seated instincts. But perhaps the disappointed tone implicit in this judgment is itself the problem. Human beings are an animal species, one that kills its own kind, one whose constituent members naturally favor short-term pleasures over the long-term interests of the species, despite being remarkably creative in dreaming of utopian alternatives. Is morality any more than froth on a daydream?

A Thoroughgoing Naturalism

What would a thoroughgoing naturalism look like here? We would continue to describe the practices of modern man in the same terms used by paleo-anthropology, avoiding at every turn external normative judgments: this is how they get food, mate, communicate, provide for their own shelter and security, care for their young, and so on. And these are their ritual activities, their ceremonies, their belief systems. Millennia from now we might add: and this is why they died out, or why they were succeeded by *Homo modesto*, a smaller less aggressive human.[2] Such a naturalism would be totally unsentimental. The sixth great extinction of other species on the planet would be but a blip in the ongoing adventure of life that would eventually move forward and flourish in unexpected ways.

But surely such a naturalism harbors a deep contradiction. Its lack of sentiment is an achievement worthy of an Olympian god. Humans are indeed wholly natural beings. And yet we have introduced into nature something utterly remarkable—a reflective and symbolic capacity that we can direct toward preexisting instinctual ends, but which equally can address the bigger picture that is now emerging. The promise of Reason may well have been hijacked by its bastard son Calculation.

Wounds to the Human Psyche

As we have seen, there have been tipping points in human understanding, points marked by Copernicus, Kant, Darwin, Marx, and Freud. Freud himself called Copernicus, Darwin, and psychoanalysis "wounds to the human psyche." In each case, a certain understanding of human sovereignty has been shown to be illusory, or at least the product of covert forces. In each case, we have had to unlearn habitual topological schemas: displacing the earth from the center of things, accepting that the world as we experience it is deeply shaped by our cognition and not just out there, that we form a continuum with other nonhuman life-forms with whom we share an evolutionary history, that human history has been shaped by those who owned the means of production, and that our normal reflective grasp of ourselves is a delicate achievement masking the work of unconscious drives we would rather not acknowledge. What Nietzsche called the death of God, Foucault the death of man, and the French *soixante huitardes* more generally called the death of the Subject—all testify to this composite new dawn. If we did not thereby resurrect a discredited Enlightenment narrative, it would be tempting to call this progress. And indeed there are many—especially, but not exclusively, religious fundamentalists—who would dispute that one or other of these displacements really does constitute progress.

Apart from Darwin, these are changes, transformations, and displacements within *human* history. What I am calling geological consciousness—bearing the significance of deep time—displaces human history itself. Geological consciousness relocates human history not just within the history of life, but also within the slow history of the geological. The Anthropocene marks significant human impact on the atmosphere, the oceans, the ice caps, average temperature of the earth's surface; it is changing what we thought we could take for granted. For the most part, volcanic eruptions, earthquakes, tsunamis, and hurricanes are local phenomena. Things eventually return to normal. But the Anthropocene is global and, for all intents and purposes, irreversible.

The End of Sovereignty

For all its unprecedented material impact, the significance of the Anthropocene can perhaps be drawn within an even deeper shift in human consciousness—the end of human sovereignty with respect to nature.[3] It is not that we lack power, or that we cannot subject some piece of the natural world to our will, if only temporarily. Crop yields have risen, antibiotics save lives, and we have learned many ways to extract oil and coal from the earth, including blowing the tops of mountains. But we cannot (or do not) control the blowback consequences of these efforts—water pollution, soil impoverishment, decimation of bee populations, antibiotic resistance, and so on. What our production processes would like to think of as *externalities* come back to bite us. We cannot control, as we would like, the various ways we nourish and poison nature to serve our ends. Our desire for control is out of control.

Within Christianity, green evangelicals, distinctly a minority, have been pushing for a revision of the translation of the influential passages in Genesis that speak of man's *dominion* over nature by replacing it with *stewardship* or *creation care* (Richard Cizik). For all its limits, this is another indication of a profound shift away from what has been called ecological sovereignty.[4] We do not have a fancy name for this shift, but it is arguably as important the others. It does not mean that we humans should cease working with the natural world to meet our needs and pleasures; it means that we should jettison any sense that the natural world is there only *for us*.[5] We should discard the sense that we can do anything we like to nature because there will always be more and there will be no lasting repercussions. Bruno Latour's recent attempts to defend Lovelock's Gaia hypothesis (according to which the earth is a living organism) follow such a path, stripping out all teleology (all "political theology," as he calls it).[6] A gaiacentric perspective would undoubtedly displace an anthropocentric one.

The sovereignty question relates not just to Nature in general but to the ways in which we treat nonhuman animal life. There are still those who resist the idea of evolution for religious reasons, and who understand the biblical story of creation literally. Until I lived in the American South I did not know this was still possible for an educated public. Some schools give equal time to creation theory, reflecting a genuine anxiety about what it is to be human. There is a line that must not be crossed. Something of this anxiety seems to rear its head in Heidegger's reference to our "scarcely conceivable, abysmal bodily kinship with the beast."[7] I will not pursue here either the details of Heidegger's account or comments on it by Derrida or Agamben, nor Agamben's elaboration of the anthropological machine (the extended theoretical apparatus by which we define the human by exclusion and subjugation of the animal).[8] That we do this seems incontestable. Why we do this is an open question.

The Anthropological Machine

Whether we can break out of our carnophallogocentrism remains to be seen. There are various obvious ideological explanations for policing the human/animal boundary, any one of which may have a depth as yet unplumbed. And it is worth reminding ourselves that the scope and scale of this war on the animal almost defies the imagination. Derrida, along with Coetzee[9] and Patterson,[10] is driven to compare the full animal exploitation program to the worst genocides. Our sense of sovereignty justifies a carnivorous diet, it legitimates human territorial expansion that drives other species to extinction as merely collateral damage, and it sustains racial oppression and the oppression of women. Once subjugated classes of humans are successfully linked to animal characteristics (such as being valued for their bodies), we can more easily exploit them (a café in Montana once provided me a tablemat depicting naked women as cuts of meat). Finally, the control and subjugation of animals may be a ritual projective enactment of what civilization demands we do to our "animal" passions—especially for violence and sex. We reinforce our separation from and control over the animal part of ourselves by externalizing the drama into a relation with actual animals. We need for there to be a human/animal abyss to motivate our own self-control.

Exploring deep time, geological time, reopens the whole question of what kind of animal we are, how we are animal, how we can embrace becoming animal, whether there is a case for human exceptionalism, and so on. For brevity's sake, let me venture some orienting remarks.

Bridging the Abyss

Deconstructing the human/animal divide can take many forms:

1. Nietzsche tied morality to asceticism and the infliction of pain; inter alia, he connected all of this to an account of how the installation of memory and identity made possible a moral internalization that sickens the human animal.[11] Freud would call this the history of repression. Both Nietzsche and later Marcuse argue that this is surplus repression and that civilization will not collapse if such restrictions are eased.

2. Philosophers contest the divide by showing that every defining difference fails because we are alike where it matters. Bentham argued that both humans and animals share a capacity to suffer. And Derrida criticized Heidegger's insistence that animals do not die but merely perish because they cannot grasp their own mortality. One strand of Derrida's response is to question what grasp we humans have of our mortality. Do we really have a clear concept of our own death?

3. The divide can be softened or further subverted by focusing on the unsuspecting depth and complexity of our actual communication, engagement, and relationships with animals (Bekoff, Haraway, Hearne, Goodall).[12] This work demonstrates that the language we use to describe these connections is crucial.

4. The results of biological, and especially genetic research, demonstrate surprising proximity between humans and our closest animal neighbors. Humans share high levels of genes with most mammals, for example with: chimps 98 percent, gorillas 96 percent, and mice 92 percent. This in fact should be unsurprising given our evolutionary heritage. However these figures can mislead; the possession of genes is one thing, their arrangement another. The 2 percent gap with chimps may be critical. Quantitative comparisons are too crude. After all, a leaky bucket is 99 percent the same as one that holds water. What evolutionary biology and neuroanatomy helpfully teach us is to recognize that some of our behaviors reflect inheritances from older creatures. Our "reptilian brain" refers to those brain structures (basal ganglia) responsible for aggression, territoriality, and ritual behavior. When we look at the continuing human willingness to go to war, we may well wonder whether the conquest, social sublimation, or diversion of "instinct" remains a top priority.[13] The first and the last of these biological ways of deconstructing the human/animal binary demonstrate constitutive dependency and are the most relevant to thinking geological time. Each of them also leaves us with more questions than answers.

I would like now to return to my cataloguing, however gesturally, the history of mankind's greatest shocks to its self-understanding. In its unavoidable materiality and unprecedented scale, we must understand the Anthropocene within this sequence, not just because of the way it marks the overflowing of human history into the geological, but because the sequence intimately connects with global warming and the Sixth Extinction (Holocene).[14] These connections should make it impossible for us to treat the Anthropocene as mere spectators.

Finally, as a chapter in this "end of ecological sovereignty" story, we must mention the human microbiome.[15] As mentioned, the human microbiome project (launched in 2007) reveals just how much of the human body is made of "foreign" organisms (including 20 trillion bacteria), whose balance is necessary to "our" health. This will likely spark a revolution in medicine and the broader ways we think about ourselves. But as a metastasizing metaphor to rethink the fragility, multiplicity, and permeability of self (and the primacy of cooperation over conflict with our own kind, our descendants, as well as other species), its implications may be even more powerful. Could the microbiome revolution not bring about an evolutionary twist in the story of Man that would change everything?

These displacements, old and new, each reflect dramatic shifts in human consciousness. But we must be cautious about interpreting this claim, especially today. These dramatic shifts in human consciousness may initially be spread thinly or unevenly. And a new grasp of the cooperative interconnectedness of life (and the folly of the sovereignty model) may coexist indefinitely with the productive/ destructive practices of yesteryear. If, as seems clear, those practices are driven by economic forces whose raison d'être remains intact, an alternative (and even widespread) dissenting consciousness may have little more impact than rain on the palace windows.

Who Do We Think We Are?

The history of our self-understanding has been a roller coaster of traumatic displacements. We seem unequipped to deal with our current situation; environmental engagement has been plagued by a sense of futility. The facts are in, but we cannot decide what to do. One explanation could be derived from Hume: "Reason is and ought only to be a slave to the passions." But this raises the question: which passions?—which morphs into how reason can best do justice to those passions. And are there not other passions than despair?[16]

What could it mean to be geologically human? At first blush it addresses the question of whether there is an adequate (or appropriate) response (or type of response) to our increasingly cosmic grasp of our terrestrial situation. That grasp has many facets: from science (new information, new paradigms, and new pressing urgencies) to new (and renewed old) *responses*. Science birthed Copernicus, Darwin, and Freud, but also Georges Lemaître (1928) of the Big Bang theory and Watson and Crick of DNA/genetics, all increasing the rich historical understanding of our "place" in nature. We have come to see our species as a product of extraordinary natural forces, operating at many levels, over countless millennia, continuous (and yet not completely) with the rest of life. We have constructed stories about these events, even as the timescales involved boggle the imagination. Most importantly we have invented new paradigms to understand the times we live in, paradigms that take us out of human history and into the geological, especially the history of nature.

Associated with these new paradigms there is a growing sense of the very real possibility of what we might now call the End of Man. We know this will happen literally on or before 3.5 billion years down the road when the sun is scheduled to swallow the earth in a fiery conflagration.[17] But somewhat sooner, a dramatic change in the fortunes of the human species is on the cards, perhaps within the century, which could certainly spell the end of a certain dream or hope—such as enlightenment, justice, peace, or generalized flourishing. Mass migration, displacement, catastrophic weather changes, disease, crop failure, starvation are all predicted. All on the heels of electric cars, better iPhones, gene therapy, and so on.

Are We One (as a Species)?

I earlier entered a caveat about what one might call our species-identity: "While acknowledging the varying impacts of different farming methods, urban life, indigenous lifestyles, hunting practices and so on, we humans are responsible. . . ." But do my efforts to capture a geological shift in human consciousness not beg the question of the unity of the human? Am I not privileging the voice of the privileged, a representative of the very race, class and economic advantage that has brought us the problem in the first place?

Who am I to say *we*, to speak for *Homo sapiens*? Is it not disingenuous to invite only select friends to dinner but include everyone when it comes to paying the bill?

This book is indeed in many ways an attempt at the performative constitution of a species-subject, demonstrating what that looks like by the kinds of claims and formulations it licenses. My defense of such a stance against the charge of abstract and misleading universalism takes two paths: first that comparative cultural studies suggests that some sort of cosmic awareness is universally distributed, as evidenced by various origin myths, and celestial observance practices, and second that globalization (for better or worse) is increasingly giving access to information about those across the seas, even to those who did not know there were seas. Species-consciousness is our destiny (as a species), whatever shape it takes (cooperation, conflict, and so on). It may well be a privilege (or a burden) to know this, but the same is true of arithmetic, speaking foreign languages, and knowing the boiling point of water. The real issue is what we suppose such privilege licenses us to say. I am working from the assumption that climate change will impact "us" as a species, not just as individuals, classes, races, genders, and nations. How it impacts us will depend both on these differential factors, on the political decisions (such as international agreements) we arrive at to mitigate their inequitable consequences, and on unpredictable events that may cut through all our categories—plague, monster hurricanes, inundations of cities, and so on.

While we are in some ways "many," there are important respects in which we share a common fate as a species; the ways in which this pans out will be influenced by how, and how far, in our political institutions, we come to embrace that common fate and accountability.

Geological consciousness is inseparable from what we have long called cosmic consciousness, which itself takes various forms—from a sense of infinite connectedness to awe when looking up at the stars. And yet there has been a change from a dialectic of oneness and distance (characteristic of cosmic consciousness) to the more geological anxiety of accountability, each operating under the umbrella of finitude. It is here that we need to turn to Hume and our broader question of the passions.

THREE

Cosmic Passions

Wonder is the first of all the passions.

—RENE DESCARTES, *The Passions of the Soul*

Reason is and ought only to be the slave of the passions, and can never pretend to any other office than to serve and obey them.

—DAVID HUME, *A Treatise of Human Nature*

In the context of a common despair about the prospect of doing anything much about climate change, we cited Hume's declaration that "Reason is and ought only to be the slave of the passions."[1] It moves us forward in trying to fathom why the overwhelming evidence of looming disaster fails to move us. But Hume's diagnosis also raises the question: From which passions should reason receive its instructions? Recall that for Heidegger we are always in a mood, and if there is any commonality between mood and passion, then the question "Which passion?" or "Which mood" becomes most salient. The role of the passions here is vital, if only because of the rightfully magnetic power of the everyday—the space in which we live, move, and earn our living.[2]

From Wonder to Angst

If there were such a thing as a geophenomenology, a serious acknowledgment of the constitutive power of such passions would be part of its remit. We have ample opportunities for awe today when it comes to climate change: hurricanes, record temperatures, mass extinctions. And the form it most often takes is angst (which easily metastasizes into despair). First, however, we should acknowledge some of the passions that bond us to the earth, indeed the cosmos, in the first place: wonder, curiosity, and delight.

Wonder

Wonder comes in many stripes. It is not confined to gazing at glistening spider webs in the morning dew. It has long operated at a cosmic register. Leibniz famously asked, "Why does anything exist rather than nothing?"[3] But this question surely has its roots in the interrogative experience of wonder. Socrates called wonder "the only beginning of philosophy"[4] and Descartes "the first of all the passions."[5] Unlike Plato, Descartes thought of wonder as a gateway or stimulus to science, which, much like Wittgenstein's ladder, could be set aside after serving its purpose.[6] This is perhaps too harsh a verdict—wonder need not impede science. Moreover, if wonder is independent of the geological, it takes on a distinct flavor when filled out geologically. The Big Bang and its primal fire has been described as "an explosion of energy powerful enough to send all matter flying apart for billions of years into the future . . . a fire that is a billion trillion times hotter than the center of the sun." This sounds like a child who has discovered the giddiness of big numbers.[7] And yet it is giving voice to wonder, running up against the limits of language and imagination, while still needing to press words into service. Nonetheless, wonder is capable of more than this dislocating power. It can be more than Nietzsche's lion that can destroy but not create value.[8] It may be just such a passion as wonder that can both break with the everyday, and yet stick around subsequently to infuse the space it opens onto with coherence and significance.

To mobilize such ongoing infusive power, Irigaray tries to redirect cosmic wonder toward wonder at the sexual other.[9] If I had not had the jaw-dropping experience of staring into the Grand Canyon at the age of ten—gobsmacked at the exposed sedimentary strata a mile deep carved by the Colorado River before there was any Colorado, over hundreds of thousands of years—I might have been seduced by her suggestion. But I was bitten with a geological wonder that I could not reassign. The American Southwest, the snow-covered Alps, and the fairy chimneys of Cappadocia affect most of us in a similar way.

Whether of the geological kind or not, it is important that wonder not be thought of as posing a question that can be fully answered. One can indeed come to delight in the exquisite mechanisms by which a specific wonder of nature actually materializes. But true wonder is fractal; each successive explanation occasions a new opening for wonder. If the very existence of the universe is explained (away), say by a divine creation story, some will feel a sense of loss—the loss of the question—even as others appreciate the reassurance of a narrative.

If preventing catastrophic climate change calls for a dramatic reversal of our carbon habits, it might be thought that its suspension of the problem-solving response might make wonder at best an unreliable ally. Those who contemplate one more cycle of mass extinction may do so with a certain detached wonder at the ways of Gaia. Will there not be as much to wonder at even if things take a turn for the "worse"? The existence of the universe will not be threatened at all. One might even come to a sharper sense of that metaphysical moment, that miraculous *event*.

At this point, it is tempting to argue that if wonder is a distinctive human achievement/capacity/disposition, we cannot both value wonder positively and remain neutral at the prospect of human decline or extinction. We would be sawing off the branch we are sitting on. But if wonder were a limit experience, an experience that brought us face-to-face with the puzzle of our very existence, perhaps beings capable of contemplating (dispassionately?) their own disappearance could feel wonder. Might not wonder be a species of the uncanny, arguably a condition for authentic dwelling?[10] At any rate, I am not convinced that the significance of at least *cosmic* wonder for motivating resistance to climate change can be saved in this way.

As much as geological responsibility requires an urgent response to the prospect of catastrophic climate change, we cannot set aside deeply disturbing questions—who "we" are (see Chapter 2), "to whom or what" we are responsible, whether we can say "my species, right or wrong," and so on. How we answer these questions will affect the shape and direction of our response. If nothing else, wonder shelters a space in which these questions can be raised.

Curiosity

Is this really a passion? Heidegger comes down hard on curiosity—along with ambiguity and idle talk[11]—for its superficial lack of engagement. But that is too harsh. Perhaps I am speaking of "deep curiosity" rather than Heidegger's target: endless flitting. I am pointing to the passion for understanding, and for Truth, fashionable doubts notwithstanding.

Consider this: I find a spearhead at the end of my field, somewhere between 1000 and 10,000 years old. I want to *know* all about it: which Native American tribe made it, how they attached it to the spear, what they hunted, what they thought of when they looked up at the stars, what this land looked like then, and so on. I walk on and find a large slab of limestone on which are etched coral fossils from about 450 million years ago, over 200 million years *before* the dinosaurs. I am told this was once a shallow sea; what would it have been like to be here then? What else was flourishing? I want to know!

Science takes off at this point, and it is just something like this curiosity that can protect science against being drawn into the economy of control and exploitation, or an offshoot of the will to power. How things work, how they arose, how they interconnect—the passion for knowledge is for many unquenchable, and not amenable to Nietzsche's reductive analysis. When it comes to the earth and its sciences—from geology and geography to paleontology and archaeology—this thirst explores origins and options; it challenges our imagination and opens ever more questions. In particular it keeps nudging up against questions of meaning. As local intelligibility ripples out into expanding circles of successful explanation, curiosity never quite sheds (and sometimes intensifies) the most significant question: Why? The earth sciences continually test this edge—focused on the earth, exposed to the cosmos, giving substance to wonder.

In saying this, I am treating Heidegger's claim that "science does not think" with a certain caution. What he means by science is the rule-governed application of concepts to the world. Narrowly following rules rules out thinking. Equally, I am keeping my distance from any reduction of science to the technological control of nature, even if it is without doubt typically pressed into service in this way. I am imagining a practice of science that remains open to the curiosity with which it began—and to our delight in our sensuous surroundings.

Delight

Consider the paintings of birds by nineteenth-century naturalist John James Audubon. They are not just accurate, but works of delight at the variety of the natural world. They capture birds in poses reflecting their trademark activities: songbirds singing, woodpeckers pecking, hummingbirds hovering, and so on. Attention to detail is foremost, rather than focusing on limits and deep meaning. Delight suspends such issues. Indeed, one could treat such careful attention as an affirmative repost to the aporetic turmoil of the big questions, while preserving the wonder that scientific curiosity risks evaporating. Delight is not disinterested, but, as with curiosity, its interest has nothing to do with control, and everything to do with respect for sparkling difference.

Wonder, curiosity, and delight are passions that bind us to the cosmos, from the largest to the smallest scale. While we cannot exclude their perverse instantiations, they typically protect us against instrumental misunderstandings.[12] With angst, however, despite important similarities, there is a change of register.

Angst and the Sublime

Kant understood "the dynamical sublime" as an experience in which we feel fearful without being afraid of the object that causes it. The dynamical sublime is "nature considered in an aesthetic judgment as might that has no dominion over us" such as the experience of watching a storm from the safety of a window.[13] A contemporary cultural equivalent would be watching a disaster movie in the safety of a cinema. Kant's idea is that a powerful (traumatic, overwhelming, disturbing) experience can provoke an awareness in us of the even greater power of reason.

I suggest that we think of the experience of the sublime rather as a class of various experiences of the overwhelming. This class would include, for example, Pascal's comment on looking up at the stars: "The eternal silence of these infinite spaces frightens me." It would include Heidegger's account of angst and *das Nichts*, captured by the sense that "beings as a whole are slipping away."[14] Nietzsche's account of alternative responses to the news of the eternal return is instructive here, as we saw in the last chapter: "If this thought gained possession of you, it would change you as you are, or perhaps crush you."[15]

What each of these versions of the anxious sublime share with Kant's account is a certain reflexive resonance. The experience is not just directed to its object; it also reflexively engages the subject, even to the point of discombobulation. In this respect, it shares with Kant's account the sense of a reflexive recoil, but it differs by not providing a higher reassurance.

Derrida's account of 9/11 brings this out. After making clear what a real human tragedy 9/11 was, he focuses on our inability to properly name the event, assigning it rather a number or a date. We like to think of it as an incursion, an attack on America from the "outside," in line with the topology of Reagan's Star Wars missile defense system. (It would have functioned like a virtual shield over America; inside we would be safe from our enemies.) We were told 9/11 was unforeseen and could not have been predicted. And yet, Derrida points out, the terrorists, trained in America, and led by Bin Laden (who was himself groomed by the United States in Afghanistan to fight the soviets), were largely drawn from one of our allies, Saudi Arabia, and used American airports and American planes in the attack. The neat distinction between inside and outside is seriously blurred. Was it unpredictable? Well, apparently it was repeatedly predicted by security analysts. Our sense of security, a fantasy to begin with, was indeed shattered. But the difficulty of naming 9/11, and the insistence that it be referred to on every public occasion, reflected the chaos it wreaked on our categories, our security at a deeper level. For Derrida, however, this is just the beginning. Linked to the breach in our sense of safety was what 9/11 portended. If *that* could happen, what might the future hold? We would be foolish to underestimate the terrorists' imaginations, or our own conjuring capacities.

This is surely a concrete case of angst, of things as a whole slipping away. For many observers there was a deep sense that the reactive nature of the US response, going to war against a constructed enemy, was an even greater source of unease (not to say despair). The terrorists really had been successful in choreographing our place in their drama. They displayed writ large the strategic incompetence of our leadership, the folly of our failure to redefine the space of engagement and so on. Sadly, such a reactive response by the United States was indeed predictable, and predicted, and its tragedy far worse than the original event. To put it crudely, it would not be far-fetched to conclude (after Ruanda, after Hurricane Katrina, after Syria, even after Puerto Rico) that the West is intellectually, technically, and morally ill equipped to deal with the kind of world-shattering disaster that climate change promises. And, going forward, it is not clear how this might change.

Pascal's remark about the stars may be thought to be voicing cosmic angst. It is a confession that our understanding has encountered something it cannot ignore but cannot handle. What we might call geological angst—a common response to grasping the significance of the Sixth Extinction, the Anthropocene, or global warming—is not dissimilar.

But there is more going on here. The gap between acknowledging a powerful (perhaps overwhelmingly powerful) phenomenon and being able to deal with it intellectually, is one thing. Understandably, so to speak, one could be drawn to religious consolation, humbly admitting one's finitude. But geological angst has another dimension, one brought out by Derrida's account of 9/11. As cartoon character Pogo famously put it, "We have met the enemy and he is us!" For those with an irredeemably upbeat sense of man's mission on Earth, this is seriously disconcerting, and some simply go into denial. But for those with a more chastened vision, the angst comes from the agonizing gap between comprehension and agency. We understand ourselves to be responsible (a potent source of guilt). Yet either we believe that it is too late to prevent catastrophe (whether for "us," our descendants, or for much of the rest of life) or we see *exactly* how current trends can be reversed but we cannot see how those solutions could be politically, economically, or socially implemented in current circumstances. Each of these responses animates agency-driven angst.

What Is to Be Done?

Much reaction to Al Gore's movie *An Inconvenient Truth* (2006) was of the Nietzschean gnashing of teeth variety. Nietzsche's scenario is oddly comforting. It *seems* that the possibility of ecstatic affirmation is at hand, up to me, if I can find the right attitude. But geologically, the angst just piles up. I would be willing to change, but it is pointless unless most others do too. And the problems are structural, endemic.

Dale Jamieson's *Reason in a Dark Time* (2014) is subtitled "Why the Struggle Against Climate Change Failed—and What It Means for our Future."[16] He writes: "Sometimes people hear this as pessimism. I say it is realism." He tries hard to guide us toward a certain optimism, but in the end this brilliant book reads as a brave level-headed struggle to pluck meaning out of the ruins of what might have been, with despair always lurking just around the corner. It will, I hope, be clearer why I am focusing on the affective.[17] The most common responses at this point are resignation, token activism (peeing in the shower to save water), and/or quiet despair.[18] We are perhaps talking about what Heidegger would call mood—deep ways of shaping and coloring our being-in-the-world. And if the immediate provocation is the recognition of our geological responsibility, there is a rippling-out effect that is equally disturbing. We realize that it is not a matter of individual agency—after all, neither I am nor any particular person is responsible for the Industrial Revolution, or carbon capitalism, or the commodification of the Earth, or *Machenschaft*. When we turn to the possibility of collective action (or collective response), hope dims more broadly, both because it is clear that democratic decision making has been fatally compromised by other interests, and because there is no guarantee at all that a true democracy (whatever that means) would make better decisions. After all, the people in Nietzsche's market square wanted the entertainer, the tightrope walker, not the prophet Zarathustra.

All this is to say that the space that unfurls when we expand Kant's dynamical sublime, or give substance to Heidegger's concept of angst, is much more complex than sitting safely behind a window watching a storm rage outside. The reflexive dimension, through which the experience turns back on us, can come to preoccupy us. If at times this turns into dispiriting irresolution (such as pessimism), the true hope of angst is that it does not get cathected (as with fear) onto a token object but keeps open the possibility of other responses. I return to this.

I am taking our passions seriously because, once aroused, they can block or inspire the change we need. And reflection on their underlying conditions can direct attention to which more basic problems need to be addressed. Nietzsche's account of *ressentiment* is a case in point.[19]

What then of Hume's claim that reason is and ought only to be the slave of the passions? Well, it's just not that simple, especially when our passions conflict, and when they are intimately bound up with what reason concludes from trying dispassionately to make sense of the world. Moreover, sometimes our passions need reeducating, or redirecting.

It might be thought to be a mark of desperation that I turned to angst for hope, for keeping possibilities open. I am tapping into Heidegger's strange but persuasive idea that, to be truly at home in the world, we must experience something of the Uncanny, the *unheimlich*.

Distance and Domesticity

Dwelling is not just about living alongside the things around us, but about cultivating a benign and intimate distance. Wonder, curiosity, and delight are not the only passions that bind us in this way to the world, but they cement and deepen our ties. Wonder in particular has an original cosmic dimension that geological consciousness only expands.

And yet passions are entangled with food and fellowship, love and friendship, comfort and recognition, where countless other passions flourish. When we contemplate the role of the passions in encouraging or inhibiting a concerted response to climate change, we also have to acknowledge those passions that arise within and sustain our nearer-term, domestic, and everyday concerns (which are frequently pressing and sometimes desperate). It is not just that we feel resignation, powerlessness, and despair. We are also trying to put food on the table, pay the bills, keep up with work, as well as enjoying a well-earned vacation, relaxing at the movies, and playing with the dog. The climate change edition of the dark night of the soul meets the trials and pleasures of the day, which often demand our immediate attention. Even if it means we have nothing to plant in the spring, we must eat the seed corn now if we are starving. And the renunciation of pleasure today in the face of reality twenty years from now is a strategy doomed to failure.

False Desires

Taking seriously what compels our short-term attention rebalances any sense that it is just despair or a sense of powerless that breeds inaction. Even if passions seem clearly tied to the individuals who experience them, a critical evaluation reminds us that they are often inseparable from the institutions that sustain them. If we then deplore the consumerist capturing of our desires, such that we come to want what we believe (sometimes, sadly, correctly) will win us the love or respect of our peers, this diagnosis also offers hope. It suggests that, at least in principle, these "false" desires could be productively redirected. And we do not need to subscribe to a technical Stoic account of a false desire to take seriously those many studies of happiness that teach us that much of what we think we want misses the mark.[20] (We think we want money, fame, and stuff. We actually want security, friends, community.) The importance of these studies is that they point in the direction of sources of happiness that are both less carbon-hungry and likelier to make us happy. The commutability of happiness on which these thoughts rest shares something with Hume's willingness to speak of "pleasures" as if they were substitutable. But the Stoic story of false desires brings a critical dimension that supersedes Hume. And when we are told that it's the narrative we tell ourselves that matters, the picture is yet more complicated. Narratives provide meaning frames for what we call our happiness, which itself arguably regulates what we experience as pleasurable. The challenge will surely be to find narratives that allow us to take on board the realities of climate change, while enriching our everyday lives. We have argued here that to respond to the challenge of climate change we need to be *moved*, that something like Heidegger's sense of the *unheimlich* helps us break with business as usual, and that wonder, curiosity, and delight foster this sense. We do not need the subordination of reason to passion, but the critical management of the full palette of our passions. They play an indispensable role both in binding us to the cosmos, reminding us of what matters, as well as impeding our response to the challenges that face us.

For there is no doubting the weight of history that descends when we begin to think geologically. We seem to be faced by layers of intractability that threaten our ability to move forward in a life-affirming way. This itself could be deadly. Nietzsche was faced with just this problem, at the level of *human* history when he wrote his early essay *The Uses and Disadvantages of History for Life* (1874). In the next chapter I explore how we might adapt his creative suggestions to a geological timescale in surprising ways.

FOUR

Thinking Geologically after Nietzsche

> Since we are the outcome of earlier generations, we are also the outcome
> of their aberrations, passions and errors, and indeed of their crimes; it is
> not possible to free oneself from this chain. If we condemn these
> aberrations and regard ourselves as free of them, this does not alter the
> fact that we originate in them.
>
> —FRIEDRICH NIETZSCHE, *Untimely Meditations*

> [Group exercises] helped us remember our bio-ecological history, as our
> species and its forbears evolved through four and a half billion years of
> this planets history.
>
> —PAT FLEMING AND JOANNA MACY, *"The Council of All Beings"*

Being geologically human is the name of a question: What could it mean to
understand ourselves as the products of historical, evolutionary and geological processes? A lot hangs on how we take up the past, who "we" take ourselves to be, and how expansive a view we have of the past. The geological
takes this question to the limit.[1]

In 1874, Nietzsche wrote an extraordinary essay, "The Uses and Disadvantages of History for Life." It was provoked by his sense, perhaps a pre-
vailing sense at the time, that a surfeit of historical knowledge could crush
the human spirit. Its failures would weigh upon us like coffins returned from
the war zone. If things looked bad then, how much worse they are now!
Continued enthusiasm for life depends on a certain capacity for active forgetting. We can and must learn from history in various ways—monumental
(drawing inspiration), antiquarian (preserving continuity), and critical (reflective engagement). Each relation to the past is essential, but taken sepa-
rately it can get out of control.

Nietzsche's discussion here is a prime example of how to nuance the relation between reason and passion, critique and the burden of the past. After reminding us of our dark history (see the epitaph), he continues,

> The best we can do is to confront our inherited and hereditary nature with our knowledge of it and [this is the critical bit] through a new stern discipline combat our inborn heritage and implant in ourselves a new habit, a new instinct, a second nature, so that our first nature withers away.[2]

Here we could speak of reeducating the passions. He continues:

> It is an attempt to give oneself, as it were a posteriori, a past in which one would like to originate in opposition to that in which one did originate. . . . What happens all too often is that we know the good but we do not do it. . . . But here and there a victory is nonetheless achieved.

This is remarkable. Nietzsche is embracing a version of Plato's noble lie—this time an encouraging narrative. But who is this "we"? He seems untroubled by drawing strong parallels between individuals and nations.[3]

> Every man and every nation requires, in accordance with its goals, energies and needs a certain kind of knowledge of the past [which] has at all times been desired only in the service of the future and the present and not for the weakening of the present or for depriving a vigorous future of its roots.[4]

Elsewhere he writes of the "natural relationship of an age, a culture, a nation with its history."[5] Nietzsche does not say that appreciating the past in the service of the future is the most *responsible* thing to do. That would be the task of what he calls the camel, the weight-bearing spirit. It is precisely the *burden* of history that needs to be lifted—perhaps through a different sense of responsibility not based on guilt. Our appreciation of history must serve "the ends of life." Great deeds often require great forgetting.

The word "life," when Nietzsche uses it, plays up and down the scale—from the individual level, through a nation, to an age—with an emphasis on the nation. Throughout the essay he describes the positive and negative aspects of each of the three ways of taking history seriously (monumental, antiquarian, and critical) showing how each can become cultural kudzu when unchecked. His deep sense is that becoming must not be allowed to be ruled by being, that Heraclitean flux not be held back by the dispiritingly congealed past. While it can be salutary to learn from the errors of the past, the past we revere was often made by those lacking historical knowledge, for whom ignorance was bliss. Nietzsche is talking about human history, in the sense of the history of a nation culture, especially late nineteenth-century Germany. And one could not imagine a more poignantly relevant circumstance for these issues and tensions to be played out than post-Holocaust Germany a few decades later.

But how elastic is the "we" here? Can the scope of these considerations be transposed from national history to broader human history, including the very emergence of the human? Could history be taken back onto the geological plane, before life itself? Could these issues be raised as a species, rather than as a nation? And when Nietzsche, in the title and within the text, speaks of Life, could we take him at his word—exploring the history of all *Life*! Or must he be understood to mean human life? Confronting our "hereditary nature with our knowledge of it" (so as to grow a second nature, better instincts) surely calls out for at least a broadening to the level of the species.

Whether Nietzsche meant this, we can also ask: how might *we* construe these remarks? Might they not equally apply to "our" deeper history? Who are "we" anyway? What if that were the issue? In case we were to think that we were betraying Nietzsche, there is reason to think that he would not protest too much to this broadening. First, Nietzsche is already talking suprahistorically, as if looking in on history from outside. This would suggest that his remarks could be broadened because we are no less inside than outside our species history. Moreover, from *The Genealogy of Morals* it is clear Nietzsche has no hesitation in transcending nationality into a speculative history of western man. And from his early essay on *Truth* it is clear he has no privileged place allotted for the human. The gnat, too, thinks it's the center of the universe. (I might add that he is himself echoing Kant's comparison of man with a self-important louse![6]) Second, Nietzsche is a materialist, or at least the inspiration for many new materialists. He distrusts every move to transcendence. Be true to the earth! Finally, it is clear he is equally interested in posthuman possibilities, as when he describes man as a bridge between ape and *Übermensch*. So what would new instincts or habits look like?

Monumental History

If we were to add his own essay to the compost heap, how might we begin to translate his essay from the historical to the geological? Consider first the monumental approach. Here the idea is that history can provide potentially inspiring examples of human achievement. If we extract from this the idea of the exemplary, and being able as we are to imagine exemplars on the grand scale, the geological offers many challenging models. A year before 9/11, the Project for the New American Century folk said we would need a new Pearl Harbor for public attitudes to be receptive enough to the kinds of change they thought were needed. Monumental history would look to events in the past that could, in various different ways, inspire geological scale transformation. These would be both positive and negative, human and natural. In human events, the social transformations that led to such milestones as the French Revolution, the abolition of slavery, the formation of the United Nations, the provision in some countries of universal health care, putting a man on the Moon—can be set against the Holocaust, Hiroshima, Guantanamo, and the Rwanda genocide. By contrast, Nature's monumental events here are largely negative——the Lisbon and San Francisco earthquakes, Hurricane Katrina, the last Ice Age, earlier mass species extinctions, offset perhaps by the Big Bang, itself as an event so momentous that the word "positive" hardly seems appropriate. Ruminating on these events reminds us of the possibility of radical social transformation, unprecedented technological achievements, the unspeakable evil men can do, our human ability to mobilize in the face of natural disaster (and our frequent unpreparedness), and the overwhelming power of nature. In this sense, the monumental can be sobering as well as inspiring. Drawing appropriate conclusions from these events and achievements undoubtedly calls for judgment, and hence partnership with the critical approach to history.

Antiquarian History

The antiquarian approach is dedicated to preserving values from the past. Here I focus broadly on what we may still call enlightenment values (values that led to the promotion of universal suffrage, education, health care, human rights, and so on). When even Derrida champions justice, cosmopolitanism, and a democracy-to-come, it becomes apparent that enlightenment values, as indeed Habermas too affirms, are being refined by further critique, that is to say more enlightenment. How would the preservation and extension of enlightenment values happen geologically? There are two obvious dimensions in which this could occur—a paradigm-shifting inclusion of the interests of most nonhumans, and of future human and nonhuman generations. This would begin to realize a more literal understanding of the scope of Life in the title of Nietzsche's essay. But for this we need to move beyond the monumental and antiquarian to the third approach to history, the critical.

Critical History

Recall Nietzsche, who said that we must not pretend the past is not full of crimes: "the best we can do is to confront our inherited and hereditary nature with our knowledge of it and through a new stern discipline combat our inborn heritage and implant in ourselves a new habit, a new instinct."

With these words Nietzsche encourages quite an agenda for the critical approach to history, and one with expansive implications. Here is a quick take on what this involves.

An Expansion of Scope

The introduction of the idea of the Anthropocene,[7] as we have earlier remarked, is a recognition of the geological scale impact of human beings, or at least the latest variation of the human, *Homo sapiens*, on the planet. The key dimensions of this impact are two: the dramatic displacement and extinction of countless nonhuman species and an equally dramatic increase in atmospheric carbon dioxide and other greenhouse gases, leading to global warming, with potentially catastrophic and irreversible consequences.

To confront "our inherited and hereditary nature with our knowledge of it" requires us to scrutinize a few things: (1) the very idea of an inherited nature, (2) the colonization of biological givens (desire for security, pleasure, power) and culturally acquired add-ons (new commodified desires) by the interests of fossil-fuel companies, and (3) our acquiescence in the face of a military-industrial complex, driven and enabled by an economic system unsustainably predicated on the externalizing of its environmental costs to finite natural sinks (like the atmosphere and the oceans). It is not clear that "we" can entirely separate ourselves—our imagination, practices, bodies, desires, values, and reflective powers—from these newly minted consumer subjects. Geological angst keeps us awake at night (or rocks us soundly to sleep), while critical history plots what Nietzsche calls new habits, new instincts.

What can we *critically* learn from an expanded "geological" sense of history? We perhaps dream of moving on from the past, letting the dead bury the dead, and for good reason. In a recent book, for example, Roxanne Dunbar-Ortiz paints a gory story of the genocidal foundations of the United States, making the fact that Columbus Day is a national holiday into an obscenity.[8] I am writing on a large farm in Tennessee-land that may well have been appropriated from the Cherokee. The Trail of Tears, an often deadly forced march of Native Americans to Oklahoma, came by here.[9] This is a past from which we (and I) are still profiting. Australian settlers decimated the aboriginal peoples they found living there—four hundred distinct aboriginal peoples, descendants of migrants from Africa, who had lived sustainably on the continent for some 75,000 years. The United Kingdom has no less burdensome a past.

We are inextricably entangled with the past. It is hard not to feel guilt. And yet guilt itself can be (but need not be) unproductive penance. As Beckett would write: "We can't go on. We must go on." We need to move on, saying never again. More to the point, while we must not forget the specifics, we must not let that Never Again be fixated on those specifics. The deep Never Again has to do with our human propensity to numbing violence, with breathtaking negligence and sheer folly, which continues today. We need to hope that the lessons we should have learned from our more than troubled past might be transferred to our current predicament.

Expansion of Critique: Complications

One thing we can learn is that the categories by which we partition the difference between human and natural events may no longer be reliable. Global warming is an anthropogenic natural event of a geological order. Some natural and human events are reversible, and we can recover from them. Even after genocide life goes on for many, even if dreadfully scarred. Deep background conditions need not have much changed. Even the horrors of war and famine can get absorbed into the historically normal. But climate change will not be like that. If it changes background conditions that substantially affect the space of the possible—say the course of evolution, the habitability of large parts of the globe, the services of pollinating insects, just to name a few—then we will learn the wrong lesson from history if we take even the epoch-making monumental historical events as our models. For the accelerating combustion of fossil fuels—the stored energy of millions of years of sunshine in a few short centuries—is a dramatic unrepeatable and irreversible "event," and as geological as they come. In response, now that the horse has escaped from the stable, it is understandable that some are promoting adaptation over prevention. For if we take the likely meteor impact that set off the last great species extinction as our model, we have no human response to learn from. The best examples we can find would probably be mass mobilization in Europe in World War II, in which all material and human resources were "mobilized" toward the "war effort."

The scope and seriousness of the response justifies reference to the war on drugs, the war on poverty, and the war on terror. But these are pale imitations of the real thing. If in the last case there is some semblance of mobilizing all available resources, the lesson is profoundly unsettling. Recent mass mobilizations have licensed the creation of a security state, with accompanying attacks on personal freedom. These metaphorical wars do offer something new, which is profoundly important for our geological reality. They introduce the idea of a mortal struggle without an identifiable enemy. Or a struggle which is as much a struggle with the webs in which we have entangled ourselves as it is with something truly external—the space of what Derrida calls the autoimmune. Again, the lesson is not obviously learned. For whatever reason, the war on terror turned into an alibi for pursuing existing military objectives (Iraq), or a hunt for specific terrorists (Bin Laden)—no way to "drain the swamp," which would attack the conditions that generate the problem, not the symptoms. In principle, such a critical approach to monumental history could serve us well when thinking geologically.

You would think that the kind of disaster preparedness inaugurated, for example, by the Katrina fiasco would be helpful. But again, previous disasters may teach the wrong lesson. For most disaster planning assumes a local problem that can be addressed by rushing supplies and people from another safe, unaffected area. But cascading global crises might not be able to be addressed like this because everywhere and everyone are affected. It is an analog to the crisis in "away" that arises when the places where we throw things "away" are full; the pieces can no longer just be moved around the board.

Jared Diamond's *Collapse* is a prime candidate for the critical approach, again with an ambivalent lesson. A critic once summarized, "His central proposition is that wherever these globally disparate societies failed the chief cause had been anthropogenic ecological devastation, especially deforestation, imposed on ecosystems of limited resources."[10] Diamond references the Ancestral Puebloan people of New Mexico, the Maya, Greenland, Rwanda, and his own state of Montana, among others. The lesson these cases teach us is that these collapses were self-inflicted, and could have been avoided if the civilizations in question had kept in touch with their ecological conditions of possibility. They made bad choices. The deeply shocking thought is that some such fate might befall not a particular local culture, but the globally unified culture of the human. While there is no precedent, how it could happen could hardly be clearer. We can respond with resignation, or apply to join a select gated underground community, or with a determination that it not happen to us, armed with the thought that it so easily could.

Real Apocalypse

Climate change can take on a genuinely apocalyptic character, depending on how far into the future we peer, and how much we focus on worst-case scenarios. But as such we might think it ought to be studied sociologically, rather than scientifically. There is a long history of failed shared doomsday fears, from that of nuclear Armageddon (widely feared in the '60s) to the Y2K or Millennium Bug (as the year 2000 dawned). Not to mention smaller groups of believers like Branch Davidians (1993), or those who followed the Mayan calendar, which predicted the end of the world on my birthday in

2012, or indeed the crew of Beyond the Fringe: "Will this wind be so mighty as to lay low the mountains of the earth?" Is not public hysteria a well-documented phenomenon? Would not a critical response to history make us pause before taking this latest fad seriously?

There are two obvious responses to this. First, doomsday scenarios or their softer cousins have rarely had the backing of 97 percent of climate scientists who, without all predicting disaster, attest to the fact that it is happening. Second, it is worth remembering the story of the boy who cried wolf. After so many false alarms the villagers did not believe him when the wolf actually came. False dawns do not rule out real dawns.

There is another problem that only a critical approach can bring out. The focus on monumental history addresses events worthy of a name. What we face in the emergence of the Anthropocene (or the Sixth Extinction) is not an event of the same order, but a sophisticated construction of the real, one that draws synthetic conclusions from a drawn-out process. One is reminded of the apocryphal frog in water being slowly heated up. The frog is boiled alive because there is never a sufficiently momentous "event" to occasion him to jump out, but just a series of incremental changes. A critical approach would have us look at the direction of processes, not just at history's fireworks. As has often been said, if we wait for a new Pearl Harbor, an electrifying event, it will be too late.

Paths of Transformation

In terms of initiating dramatic changes in consumer lifestyles, such momentous events (or their dramatic equivalent) may be necessary because people need to be motivated to change before they actually *have to*. And it may well raise what could be dubbed a logical problem. What if it involves *new values*? How can we judge values from the point of view of the old paradigm, where we are now, that only make sense in a new paradigm? This is the deep problem behind Nietzsche talking about new habits, new instincts. Can we enter into such new arrangements voluntarily or only under duress, under the pressure of circumstances? Here again, a question of passion. Serving what passion could reason instruct us of the need for new passions? Hobbes would say fear.[11]

One specific way in which the critical dimension to history would be important is in welcoming a wider sense of life, one that included nonhumans, and indeed nonmammals, as equal stakeholders on planet earth. I will not here offer arguments for such an extension. They largely rest on the common capacity for suffering, or for having a life worth living. It is important also to bring out the radical interdependence of the many life forms and the folly of cherry-picking the friendly furry ones for special treatment.

Affective Resonance

These arguments are not without questionable assumptions. It may well be that some sort of affective resonance, often in short supply, may be necessary for such arguments to carry in practice. But if truth be told, this is no less the case for the various other forms of ethical extensionism that push obligation beyond family and friends and people of the same sex, color, or ethnic identity. Human history gives us countless examples of progressive inclusion within ethical protection of both humans and animals, especially on racial and gender grounds. Animals and ecosystems are arguably next in line for the full treatment.[12] How radical a shift would it be for each of us to think of ourselves primarily as living beings, not sentimentally, but in *a speculative-practical way*, treating other beings, directly and indirectly, with respect? This thought is only reinforced when, as we have already mentioned, we consider the microbiome project, which teaches us that we are each already communities of creatures. The expansion of the "we" into Life itself is the conceit behind my reworking of Nietzsche's essay. History, as they say, may not be on our side, as more and more children suffer from "nature deficit disorder," and do not have the visceral confidence or intuitions to easily engage with the natural world, let alone identify with it.

The same applies to future generations—stakeholders who do not yet exist, with real interests but no voice—both humans, and entire species of nonhumans that may be wiped out.[13]

Creative Forgetting

This is not straightforwardly to endorse everything Nietzsche says. Take forgetting, for example. Forgetting, in the everyday sense, is the source of much folly, inadequacy, failure to realize our hopes and ideals. So, is there not a willful perversity in Nietzsche's championing of forgetting? As I have done at various points, we need to refresh and reimagine the significance Nietzsche gave to forgetting. Forgetting and remembering play a role in many philosophies. Plato associates philosophy with remembering (*anamnesis*). Heidegger's destruction of the history of ontology combats the forgetfulness of Being. And surely it is not without reason that we say "Never Forget" to events in our barbarous past to avoid repeating them. Or as Santayana wrote: "Those who cannot remember the past are condemned to repeat it."

But with only a little license, let me suggest some ways in which we can support positive forgetting. Pessimism about the future often rests on the anticipation of tipping points that will bring accelerated change and spell disaster. But nonlinearity, temporal discontinuity, can come to the rescue. Public opinion is subject to tipping points in the other direction—a mass Eureka experience about the urgency of climate change. We need to "let go" (forget?) our simple assumptions about linear time not least to motivate persistence in the face of the seemingly intractable. Time could surprise us!

If we really are to take seriously "new habits," "new instincts" (or new passions), we need to do more than forget some of the old ones. We need, as Freud might say, to work through them, so we can set them aside. This is a seriously difficult process for which "active forgetting" would be a good term. There are clearly issues to do with forgetting when it comes to the historic carbon privilege—polluting and profiting—enjoyed by rich nations who would be all too happy to have this forgotten by China, India, and Brazil. This needs to be acknowledged, and in that sense remembered, but not in a form that licenses more environmental negligence. The wealthy West needs to "forget" those good old days, and help develop new standards and values. On the other side, there is recurrent need for a very concrete form of forgetting: debt forgiveness, allowing poorer nations to escape poverty and debt repayment cycles in order, precisely to be able to think in the long term and make environmentally wise investments.

Affirmation of Life

It would be hard to come down against the affirmation of life, and valuing what is life-affirming. It will however rightly be said that the devil is in the details. Is it an affirmation of life to smoosh the mosquito drilling for blood on my neck? Whose life? Sartre's Roquentin "loved humanity" but had few if any friends. Don't we need to get specific about life? Do we really have any idea what we mean by life-affirmation? Or by putting history in the service of life?

In crude ways we do. Biodiversity good, monoculture bad. Flourishing good, being poisoned by toxic waste bad. A vibrant healthy ecosystem good, extinction bad. Of course, it's true that most of these terms are tendentious or contested. And if we said earth with humans good, earth without humans bad, I would want to say: Slow down; let's fill out the picture and define our terms. A critical inheritance of our deep history is a good thing precisely because it does not just answer these questions; it keeps them alive.

As for forgetting, it is clearly a matter of management or judgment, even for Nietzsche. We need to be able to forget, but at times, he wrote "[We may also need] a temporary suspension of this forgetfulness; [when we recall] how unjust the existence of anything—a privilege, a caste, a dynasty, for example—is, and how greatly this thing deserves to perish."[14]

So, what happens when we attempt to expand the scope of Nietzsche's reflections on history? Each of the paths we set out on raises questions to which there are no easy answers. How do we bridge the gap between human history and natural history? What is it to welcome, affirm, even identify with a wider sense of life than the human? How do we judge when to forget, and when not? Doubtless these issues deserve a more extended treatment than they receive here. But we have perhaps exercised these questions to the point at which it becomes clear that the challenge is not to answer them as to take them seriously. Derrida would call this "going through the undecidable."

Mobilizing Angst

It is perhaps time to resurrect Nietzsche's central concern—does not such a critical geological consciousness impose an intolerable and incapacitating burden on us? Don't we need precisely to stop taking all these questions seriously in order to flourish?

What I have called geological angst is not a choice but a condition we find ourselves in. We can convert it into fear (by constructing an object to make practical action possible), or in some other way resolve it. But I recommend instead creative ways of exercising a certain negative capability (an ability to tolerate uncertainty) to keep the angst alive. As I see it, a critical consciousness, drawing on the monumental and antiquarian approaches to history, does just that. It does not prescribe remedies, but rather, in Heidegger's language "prepares the way."

Angst and Attunement

> Thinking through affect is not just reflecting on it. It is thought taking
> the plunge, consenting to ride the wave of affect on a crest of words,
> drenched to the conceptual bone in the fineness of its spray.
>
> —BRIAN MASSUMI, *Politics of Affect*

> I'm afraid that the planet will hit us, anyway.
> Don't be . . . please. Dad said there is nothing to do, then . . .
>
> —LARS VON TRIER, *Melancholia (film)*

I have connected Nietzsche's sense of the burden of the past with the common mood of resignation about climate change. I gave substantial ways in
which we might think of our present age in terms of mood, with glances to
Heidegger. And by rereading Nietzsche's "Uses and Disadvantages of History for Life," I argued against Hume to critique the passions, with some
support from Nietzsche himself. I want now to redeem some of the promis-
sory notes I have been issuing while trying to avoid a wholly satisfactory
conclusion. It's not that we cannot get *any* satisfaction, but if we try sometimes, we get what we need. Who, we?

The declaration of the Anthropocene opens us up to a geological consciousness in which we are no longer mere spectators of cosmic time; we
recognize ourselves as participants at a geological level. This means that we
are having the kind of impact on the earth's physical systems previously
reserved for meteors, massive volcanic eruptions, and so on. Deep time is
coextensive with geological time, but operates as the horizon of questions

of meaning, self-understanding, and responsibility raised when our rearview mirror knows no obvious limit. Geological time allows us to look back to the Big Bang. Deep time refers to the issues and questions raised when we expand our historicity to a geological scale. The climate crisis gives a distinctive shape and urgency to the Anthropocene. It extends the geological scale of reference into the future and turns extinction into a live possibility for our children's children (instead of an abstract calculation of 3.5 billion years, when the sun dies). Our personal mortality is now nested within a grasp of potential catastrophe or extinction of our species. In short, if geological time is the time of the big numbers, essentially calculative, deep time is the time of geologically expanded historicity. My thought experiment is to expand Nietzsche's concerns about history to those of geological history, which takes us into the existential (and other) issues raised under the heading of deep time.

One of the real possibilities of the deep-time perspective is that we might come to see the emergence of a different kind of time, which might not be chronological at all. That happens for example, for Marxists, when the proletariat grasps the possibility of class action. I recall Lucian Goldmann's claiming that Heidegger had Lukács's *History and Class Consciousness* open on his desk when writing *Being and Time*. Heidegger and Derrida both explicitly suggest such possibilities. The to-come (*a-venir*) is no ordinary future.

Bad Moods

Heidegger seems to find hope in a change of mood: "We must awaken a fundamental attunement, then! The question immediately arises as to which attunement we are to awaken or let become wakeful in us." Confronted with the real urgency of climate change, the worry is that we fall into resignation or apathy or some such mood that would prevent any kind of response to its challenges. His discussion of boredom is a case in point of how deep a mood might be. Boredom is an attunement that is essentially tied up with time, and when it is public it has perhaps some parallels with apathy. Our foray into discussing the present age concluded provisionally that it could be plausibly represented by something like *ressentiment*. This is clearly not the whole story but it might at the very least capture a decisive sector of the population. Suppose then, for a moment, that we have to deal with two broad contemporary moods of *ressentiment* and resignation.

Nietzsche's essay sets us up to consider in his terms how we might negotiate the burden of history, transposed onto a geological scale. My account privileged critical history, and I gave a positive role to angst. This leaves us with two unanswered questions. First, how precisely can angst play the positive role I have attributed to it? Is it not both primarily individual and debilitating? And second, if attunements condition our engagement with the world, if we find ourselves mired in resignation and *ressentiment*, and if we cannot just press the mood-reset button, what can we do?

The Power of Angst

In response to the first question, my argument goes like this: Some sort of cosmic angst has a history independent of global climate change (which is how Pascal, Kant, and Heidegger can all talk about it). I am recruiting them onto one team—Pascal's fear of infinite space, Heidegger's sense of the whole slipping away, and Kant's sense of the sublime. In the case of the last two at least it is clear that angst has a reflexive power. In Kant's case it is to disclose the privilege of Reason. In Heidegger's case it brings us face to face with our being toward death, and the possibility of authentic resoluteness. I conclude from this that angst is not just a source of paralysis but also existentially reflexive. As such it can open possibility. I do not know whether a philosophical guide is necessary here, or whether the recognition of the power of angst is part of the experience itself.

I am tempted to draw a parallel with Kierkegaard's analysis of despair in *Sickness Unto Death*. There he insists that, in addition to a suicidal despair (despair at willing to be oneself), there is another form, "despairingly willing to be oneself" made possible, essentially, by faith. We are not alone; it's not all up to me. Kierkegaard is not just recording a philosophical insight but describing, indeed performing, an on-the-ground existential movement, which enables us to carry on. Analogously, I understand Heidegger to be describing a lived movement from angst to resoluteness that philosophers merely describe après coup, cutting in before the Owl of Minerva takes flight.

I do not have a definitive answer as to whether we can understand angst collectively when it first serves precisely as a moment of personal reindividualizing recovery from the anonymous grip of *das Man*. Heidegger eventually offers an account of a collective resoluteness—via struggle and communication—through which a people comes to see itself as an agent, with a common destiny.[1] Could angst play a role here? And if so, why not also when we find ourselves faced with the possibility of our demise as a species, or the earth's bounty slipping away?

Deconstructing Mood

Let me now turn to the second question: How we might be released from moods in which we seem to be mired, the dominant moods of resignation and *ressentiment*? It seems clear that we are not faced with any sort of Adornoesque negative totality, a kind of metaphysical black hole. However dominant resignation might be, it is not the only show in town. There are plenty of "optimists" (though some are misguided, and some crazy), and others willing to be persuaded that it's worth doing something while there's still time. So we do not need to invent a deus ex machina such as the "call of conscience" to break into business as usual.[2] A quick version of my argument here would be this: Moods are grounding in important ways (not least opening, closing and shaping the future). But however we slice it, moods are not themselves without grounds. Contemporary *ressentiment*, whether we see it as a continuation of a long-standing condition—slave morality—or as an ugly resurgence of a dark latent human possibility, has its roots in conquest, power, domination, exploitation, and economic stress. Its grounds, today, look to be global economic restructuring.

If that is so, then one would have to tackle *ressentiment* by tackling the economic policies that give rise to it. That then opens onto some familiar options: on the one hand, nationalism, racism, sexism, imagining one could regain some privileged place in the system, often leading to war, dictatorship, and so on. Alternatively, one could tackle the economic system itself.

I believe we are indeed faced with this choice. But matters are not simple, as the following case study makes clear.

An Interlude: The Hysterical Prohibition of Reflection

> The number one rule of US media discourse is that whenever there is violence or attacks, the one thing we don't want to do is to think about the role we played in provoking it.[3]

After the May 2017 Manchester bombing, Jeremy Corbyn, the leader of the British Labour Party, condemned the bomber unequivocally but reminded people that terrorist attacks were not unconnected to Britain's foreign policy, not least the invasion of Iraq. He had said as much in arguing against that invasion in the first place. And his remarks were in line with many military and intelligence experts. The response from his Conservative opponents—this was election season—was to accuse him of justifying terrorism. His direct condemnation was ignored by a reactive anger that elbowed out even the expression of a steely determination to get to the bottom of what had precipitated the event. His measured response to the poisoning of a former Russian spy in the UK In March 2018 received a similar reception. One-dimensional patriotism can trump critical thinking.

Tony Blair, the now widely reviled UK ex-Labour prime minister and an enemy of Corbyn, had years ago used as a slogan, "Tough on crime and tough on the causes of crime." Corbyn had in effect said the same thing about terrorism. The extra element was to remind people that we ourselves— our own actions—can precipitate unwelcome responses.[4]

After 9/11, the then mayor of New York City, Rudy Giuliani, proclaimed, "Those who practice terrorism . . . lose any right to have their cause understood." He is making a remark similar to that of Corbyn's critics. The fear is that to "understand" is halfway to sympathizing and worse, forgiveness. And that would weaken our unequivocal rejection of violence and our solidarity with the victims.

This shows the further limitation of Hume's slogan. The passion for revenge is blocking our capacity to reflect on how to prevent such occasions from arising in the first place. If a rabid dog bites you, then of course you need sympathy and treatment. But we would not normally think it inappropriate to suggest that rabies in the wildlife population also needs to be tackled. The difference perhaps is that terrorists are freely making decisions. As such they are responsible for their actions, and can properly be condemned. That is true. But why is there such an outcry against insisting that free decisions might be affected by background circumstances? Nietzsche once described the idea of free will as a "hangman's metaphysics" precisely for this reason. Exculpatory circumstances get in the way of anger and revenge. What we are up against here is the juxtaposition of (1) two moral grammars—responsibility and causality, (2) two temporalities—immediate feeling, subsequent considered judgment, and (3) two affective registers—anger and revenge (heart on fire), and cool detached determination (brain on ice).[5]

After the Manchester bombing the head of a Muslim anti-extremism charity was spat on in the streets: "You killed our children." After 9/11 "we" invaded Iraq, despite the lack of any link to the hijackers. There is said to be a connection between the adrenal gland and our fight/flight response: It's really useful to have an accelerator pedal to get out of immediate danger. And while if you are attacked in the dark it might make sense to hit out at anything, the tragedy of allowing such natural responses to be translated into responsible political discourse goes without saying. Derrida ("Philosophy in a Time of Terror") commented on the requirement to talk about 9/11 and condemn it (which he did) while going on to deconstruct the frame in which the event had been presented (an unforeseeable attack from "outside" by "them"). He ran the great risk attached to philosophizing—for which Socrates was famously punished—of failing to respect the prohibition on stepping back from anger and revenge to understand what is going on, even when the point of doing so is to prevent its repetition.

The Backstory of Resignation

If we focus on resignation on the global climate change front, we face a similar circumstance. Resignation seems to reflect various agency problems—classical problems of collective action, problems of responsibility, constituency, and many others. But what is the backstory to the *Stimmung* of resignation, to the sense that it all seems hopeless? Interestingly, given our doubts about calculation, one excellent source is Bill McKibben's *Do the Math*, both an essay in *Rolling Stone* and a nationwide tour (2012). He argued that bringing the rest of the world up to Western levels of consumption, especially of fossil fuels, would require four earths to cope with the CO_2 produced without triggering huge (potentially catastrophic) climate instabilities. The reserves of coal, oil, and gas held by the fossil fuel companies simply cannot be burned without disaster. If we are trending in that direction, those trends are literally unsustainable.

To cut a long story short, I argue that we are indeed faced with a "choice" between the Unthinkable and the Impossible. On the one hand the Unthinkable—something like Hobbes's war of all against all. A decimated population could seriously cut our gross carbon footprint! On the other hand the Impossible—what we could call an economy-to-come. Here I would fuse together what Derrida dubs the "im-possible," and Che Guevara's exquisite slogan: "Be realistic: demand the impossible!" I develop this stark choice at length in Chapter 9. I will argue that the impossible offers grounds for a certain hope, and hence a way out of resignation—while there is still time.

Aporetics of Agency

We are faced with multiple problems of agency. If "we cannot go on like this," many questions arise: Who "we" are. How individuals and groups can act today. What agency looks like. Whether action at all is what called for (perhaps thinking?). How to change or anticipate change in the very frame of action, if that is the problem? How to square different formulations of the need for radical transformation. These questions are not new: What ought I to do? (Kant), What is to be done? (Lenin) Perhaps different today, or new for us at least, is that the future is not what it was; it seems broken. And if agency presupposes projection onto a future horizon, we have a problem. Although much of what I will go on to say falls under the general aporetics of agency, I will start by exploring some of the specific cracks in the future.

I have argued elsewhere that as much as I appreciate Derrida's thinking of the to-come and an im-possible, unrepresentable futurity, we must not allow that to translate into blind complacency, an alibi for inaction.[6] I cited all those claims that 9/11 was unforeseen, despite many who predicted it. We might say the same about the Iraq war fiasco. But might not these examples just persuade us to be vigilant, to think ahead, to be less lazy? What is the big problem about the future? The classical thorny problems about the future start with the fact that it can never appear as such. It never arrives, for as soon as it does it ceases to be the future. And indeed, more mundanely, achieved goals can lose their attraction. This concern becomes real when for example the future is a source of hope, and has to remain out of reach, like the donkey's carrot. Or if the future is importantly open. Then again it may be said that while we always access it through representations, we cannot *know* anything about it in a strong sense. We have to rely on induction (see Bertrand Russell and the chickens)[7]. But this can seem a very abstract puzzle; surely that strong sense of *know* hardly ever comes into play. Thanks to Rumsfeldian epistemology, we know there are known knowns, known unknowns, and unknown unknowns. Get over it—learn to navigate.

Beginning Again, with Heidegger

Heidegger will come to speak of Dasein not as a given but as a task.[8] It is as if he has swallowed something of the challenge of Nietzsche's *Übermensch*, for he specifically speaks of this as a burden. And when we think of Nietzsche's reference to the burden of history in "Uses and Disadvantages of History for Life," this marks a dramatic difference in tone for Heidegger. Is he accepting that history might be a burden for life, but not for something beyond life, like Dasein? If we then couple this with his *Contributions* story, that of preparing the way, the transition from the First beginning to the Other beginning, we have something of a rerun of the destruction of the history of ontology begun in *Being and Time*. This time, however, there is a renewed sense of the role of language in enacting this transformation. What is quite distinctive is the painstaking meticulous even obsessive caution with which Heidegger approaches this task. If *Being and Time* had put new wine into old bottles, *Contributions* tries new bottles, new language. Performatively, he must write the change he wants to be. The jury is out on the necessity of this strategy. You do not need to be a fat man to drive oxen, as it used to be said. But that Heidegger feels driven to write quite differently does reflect the seriousness of the problem.

What Heidegger comes to call inceptive thinking would at least point toward an alternative to the first beginning—of thinking and being in the West. I understand him to be saying that the first beginning launched what he calls *Machenschaft*, calculative thinking, one that takes charge of being. We need another dispensation, one that will rather let beings be. What this entails is far from obvious. We have premonitions, adumbrations, but if we thought we could put our finger on it, we would most likely still be thinking within the orbit of the first beginning.

In *Contributions*, Heidegger might be thought to be modeling the problems with the future to be found in the very different arena of climate change. But it is at least worth asking whether there might be a closer connection. Climate change is no singular problem. There is the problem of dealing with, or mitigating, the slow catastrophe already underway; there is the problem of trying to change course before it's too late, especially if it's already too late; and there is the problem of changing what might be called our fundamental disposition, the biblically endorsed domination of nature. The first two, as we will see, are not just practical, but move in that direction. And even if it has now become an ecological cliché, this last problem is deep. For many, the domination of nature is symptomatic of a wider pathology of relationality, in which domination is central, including political, and race- and gender-based domination, and indeed an instrumental view of language and reason. Such an account slides into one of colonization and subjectification, in which the creeping constitution of the subject by the reigning discourse makes it harder to access any residual subject of domination. The word "subject" itself may lose its grip.

My point is that we cannot straightforwardly contrast Heidegger's search for an Other beginning with those trying to reimagine the world in ecological, social and political ways. It is sometimes tempting to cast Heidegger as Hegel, as the idealist, and then play materialist Marx to his Hegel. But that would reproduce simplistic and unhelpful binaries. Instead I suggest here that in what we may think of as more "material" dimensions—like agriculture or politics—we will repeatedly find a convergence between limits on the ground, and of the ground: materiality and being-historical if you like. As it happens, the Marx reference becomes more plausible when one recalls that materialism for him meant taking seriously "modes of production"—that is to say power, domination, and relationality. So, if we believe that unrestricted free market capitalism is unsustainable because it presupposes unlimited growth that will destroy the earth's systems, this account runs parallel to one that says that our need or desire for the control of nature is out of control. If the production of goods for consumption produces waste that needs to be dumped in ocean and atmospheric sinks, the production of knowledge generates simplistic oppositions, and inadequate concepts, that quickly wear thin and call for ecophenomenological refreshment.

Heidegger's quest for the Other beginning cannot simply be identified with what he might call being-geological thinking, with appropriating the burden of deep time opened up by the advent of the Anthropocene, or specifically with a transformation of our fundamental disposition toward nature and much else, from exploitation to (say) receptivity. But I invite reflection on the challenges of translating one formulation into the other, and the uses and disadvantages of pursuing these parallels. To repeat: The parallels between the formal difficulties of Heidegger's quest and thinking deep time are striking. For eco-warriors, the central problem is how to motivate changes in behavior on the part of those whose practices and passions are intimately tied to the toxic economy of growth and consumerism. How can we move from thinking outside the box to acting and wanting outside the box?

This is the same problem faced by Kierkegaard in trying to justify the move from the aesthetic to the ethical.[9] It requires a leap because the ethical makes no sense from the aesthetic standpoint. Heidegger similarly needs the call of conscience to break in to everyday complacency.[10] And for Nietzsche's Zarathustra, the people in the marketplace simply cannot hear what he is saying. The last man is deaf to news of the overman. Heidegger's solution, like Nietzsche's, is to invite the reader to share the problem.

When Derrida says that future can only be anticipated in the form of absolute danger, he is speaking of a future that breaks with business as usual.[11] In the case of 9/11, the break is a traumatic crack in our sense of security at every level, and not one that promises any brave new world. But there is even a sense of danger when, at some level, we realize that Rilke was right when he wrote, "you must change your life."[12] That is, "we cannot go on like this."

When Derrida writes of the to-come, of the im-possible, the future takes a different turn, somewhat akin to Blanchot's reference to the Messiah at the gate.[13] It is not simply that we have a break with linear temporality. Rather, what is at stake is an event within time that would inaugurate a different form of time, not just a branch in the existing road system. The Messiah would come to redeem us, let us be born again. A transformation in our being toward death would change how we lived.

Arguably the Anthropocene is such an interruptive event. It may be able to be dated within this time series, but, as Naomi Klein put it, "this changes everything." We jump rails into a different way of being in time. As an example of an as yet merely larval thought, imagine we followed Heidegger in taking being-toward-death seriously but found a way to expand the significance of our own lives/deaths into that of others, even nonhumans. Those who identify with the innocent victims of war manage this. Those like Elizabeth Costello who take on board the sixth extinction (of animals) do this.[14] And appropriately, this can take them close to what we call madness. My point is that such a diversification of finitude would precisely "change one's life," open a new orientation to time from within this one. Arguably, this is what is happening in those lines of Beckett: "I can't go on. I'll go on."

Perhaps what is distinctive about climate change is the way this sort of transformation of disposition from exploitative to receptive (which would change the shape of the projective future from growth to something else) is intertwined with the ongoing "real time" increase of CO_2 in the atmosphere as it passes various critical system–destabilizing thresholds. Timing is of the essence here.

There is much more to be said about the future, especially to do with apocalyptic scenarios, utopian dreams, and the role of prophecy and promise. The bottom line is the powerful role played by such projections of the future, even invisibly and even when one imagines oneself free of them. The significance of providence in Kant's thinking is such that the absence of God would truly be a *disaster* for him.[15] One explanation of why we are not addressing future climate catastrophe might well be that addressing it threatens the future horizon at any level we can make sense of. We can at best contemplate action within such fundamental frames. The idea that the horizon itself is in question is simply impossible to accept.

Who is this "we"? When we talk about "the future," it is easy to forget exactly who we are talking about. Many of us can safely assume that we ourselves will not get swept up by raging tsunamis, heat waves from hell, or rising sea levels. We worry about our children, about future generations. And as Sam Scheffler argues, they matter to us in ways not at all obvious.[16] It's not just that we don't want to see them hurt, but our assumption that human life will continue into the foreseeable future is a horizon that gives our own lives meaning. This shifts the question from life after death to another sense of the afterlife. And one could perhaps make the same argument about other species; we take for granted their continuance in ways that we don't realize. In the case of other humans, we perhaps project continuing-to-matter-after-our-death, or contributing to a family lineage, or a body of knowledge—being part of something ongoing. Knowing the show would cease shortly after we died would change the meaning we could give to our own lives. If this is so, the question of who "we" are becomes directly important.

Who Then Are We?

As we quoted earlier, Heidegger insists, "We must awaken a fundamental attunement, then! The question immediately arises as to which attunement we are to awaken or let become wakeful in us. An attunement that pervades us fundamentally? Who then are we? . . . And this history of spirit—is it merely a German occurrence, or is it Western, and indeed European? Or should we draw the circle even wider?"[17]

This of course was the question we posed to Nietzsche. Whose history and whose life! Can we humans, today, inherit deep time or geological history? And on behalf of whom or what? And are we talking about academics like us, with expansive and speculative sympathies? In place of an extended review of the questions here let me just list some of the most important considerations:

1. There is a "we" of constituency, a "we" of agency, a "we" of rhetoric, a "we" of impact, a "we" of concern, and a "we" of responsibility. And then there are actual and aspirational we's. These senses are all typically confused.

2. Zarathustra's search for an audience leaves him only with animals. Humans don't recognize his "we"; "I am not the mouth for these ears." Heidegger in *Contributions* seems to address only the few and the rare.[18] This "we" is the we of a common language, a way of talking, a shared view of things—not be taken for granted.
   Will there be a permanent rift between those who get it and those who just don't, those who are and those are not part of the solution?

3. If "we" cannot go on like this, are we talking about the average carbon footprint of an American—five times the global average? Or we humans? And what is it to speak on behalf of others in this way? Can I say, "We cannot go on like this" to indigenous people who have no part of the problem? Or to nations still developing?

4. Who is the we that is threatened? Those of us alive today? Those future humans who would be alive if we had room for them? Future people who can be expected to be born, projected ad infinitum?

5. From the point of view of agency—how can we get our act together? And who are those we's? Democracy seems to be failing climate change. Are these we's of collective initiative, or citizens dragged kicking and screaming into a greener world by legislation?[19]

6. What of the we's of corporate responsibility?

7. Can we represent, or care for, the interests of nonhuman species? If so, how? Under what conditions? When they are useful to us? Or for their own sakes?

8. Given the microbiome project, according to which much of you and me is bacteria, should I be saying we, even alone on a desert island?

It may be a stretch, but Heidegger's sense of the possibility of awakening or reawakening a fundamental attunement would go a long way toward a persistently expansive and generous response to these various questions. We now turn to a particularly rich example of his thinking about deep attunement in the present age—his discussion of boredom in *Fundamental Concepts of Metaphysics*.

The Present Age: A Case Study

It was the best of times, it was the worst of times . . . it was the epoch of belief, it was the epoch of incredulity, it was the season of light, it was the season of darkness, it was the spring of hope, it was the winter of despair.

—CHARLES DICKENS, *A Tale of Two Cities*

The future can only be anticipated in the form of an absolute danger.

—JACQUES DERRIDA, *Of Grammatology*

Historical periods have not waited for philosophers to be thought of affectively. The Depression of the 1930s is a case in point. On the other hand, philosophers have not only given a special privilege to certain moods—like wonder, angst, fear, despair, hope—but they have also come to understand their own times in such terms. Kant gives pride of place to courage and hope, in the face of human greed, laziness, and vanity.[1] Kierkegaard writes of an age of indolence, reflection, and the absence of passion.[2] Nietzsche highlights *ressentiment*,[3] the need for affirmation, active forgetting, and the strife between Apollonian and Dionysian. Husserl's *Crisis* is dominated by disappointment about reason, "the dream is over," and alarm at the fate of civilization. Heidegger, in *Contributions to Philosophy*, speaks first of an original wonder, and now of our plight, of shock, restraint, diffidence, presentiment, and foreboding while elaborating what he calls our basic disposition.[4] Simultaneously, he writes of "the darkening of the world, the flight of the gods, the destruction of the earth, the transformation of men into a mass, the hatred and suspicion of everything free and creative," insisting that childish words like pessimism and optimism do not begin to capture the situation.[5] And Derrida will write of the ten plagues of the New World Order, and later of a "war on pity."[6]

Heidegger has a complex relation to time and history. In *Contributions* he is still committed to a version of the *Being and Time* project of relaunching the human project by a transformative return to the Greek inauguration of philosophy. The first beginning is to give way to an Other beginning, but the transition is archaeologically painstaking. This may explain his rejection of vulgar talk of pessimism. Heidegger is concerned that his thinking not be confused with a Spengler-like *Kulturpessimismus*, for he understands the spiritual decline of the West as happening at an ontological level, not a cultural one. This is on a par with his excoriation of "lived experience," writing instead of basic disposition, *Grundstimmung* rather than passion or emotion. Part of *Contributions* is engaged in a hermeneutic of historical mood: distress, dismay, and shock, and how best to respond to this. We might think that this reflects the plight of German life in the 1930s, but Heidegger's references to Hölderlin earlier in the nineteenth century, and to the flight of the gods, suggest otherwise. It will make a simple chronological understanding of the present ever more implausible.

The Mood of the Times

This raises the question of historical mood more generally. By that I mean how to diagnose a mood, how to avoid obvious simplifications or idealizations, how to think through the relationships between a mood and its underlying conditions, what possibilities for transformation there are, how moods affect agency, and so on. And this is surely relevant to our own time. It is a crude measure, but if you trawl the media for comments about the present age—especially ISIS, Trump, climate change, and Brexit—it is hard to escape the language of passion and mood. This is typical: "We've got a lot of reasons to be angry. But the country is in a very Dark-Side-of-the-Force mood, convinced that anger is empowering, not blinding."[7] These comments lack philosophical sophistication, but they do invite reflection. Their authors casually attribute feeling to individuals, groups, or whole classes. This was clear in references to the Depression in the 1930s (after the panic of the 1929 Wall Street crash), in which economic, social, and personal dimensions were seamlessly connected. The justification for this is that many people were exposed to the same conditions, social contact reinforced individual feelings, the media can organize and channel emotion, and inhuman entities (like markets) could be influenced by human mood and register it in analogous ways.

Whatever causal mechanism is being invoked, the narrative intelligibility in which feeling leads to action (or inaction) seems appropriately unproblematic. Clearly affects may be epiphenomenal, mere accompaniments to action. And one's passions may have deeper unacknowledged springs. But that does not typically make them less real or less efficacious.

What these media scraps do not tell us is whether this condition is normal or exceptional, whether the attributions of affective motivation to this or that ground is "warranted" or just political, whether the anger, fear, anxiety, or hope is being directed at appropriate targets, and how that knowledge is distributed. Are the forces in play today the usual culprits or something new? Are they are coming together in a more global "mood"?

The public, it seems, is angry. People are said to respond to climate change with resignation, despair, and apathy. Mass immigration is leading to fear, hostility, and desperation. All of these are, of course, fodder for political manipulation. How are they connected with individual angst or dread? Or hope? Or the "American dream"? Can "we" operate individually or collectively without such affective temporal horizons? Nietzsche's sense of history as a (crushing) burden, in the face of which we need to find "positive" ways forward, even in the face of the follies and atrocities of the past, seems alive and well.

Especially in *Contributions*, Heidegger is concerned performatively to enact thinking, not just talk *about* stuff. Can we translate this into a more contemporary idiom, responding to our own crisis (an unsustainable economic system, catastrophic climate change), without betraying Heidegger?

The statements I quoted from the media might seem to lower the tone, to be precisely the *Kulturpessimismus* from which Heidegger was trying to distinguish himself. But this is a misunderstanding. Those who penned these various remarks didn't need to understand the deep significance of what they wrote. The point is: How do we interpret these claims? We can do so as expressions of personal feeling, or of collective consciousness, or as the shuddering of the tectonic plates of being.

A Philosophical Reading

As an oblique way of addressing this issue, let us take a glance at Heidegger's fascinating discussion of attunement in *Fundamental Concepts of Metaphysics*.[8]

Heidegger offers a brief survey of the popular interpretations of "the current situation." He characterizes them all as literary, cultural, or journalistic. But the constant motif of a relation between life and spirit reflects, he suggests, a common origin in Nietzsche's thought, especially the tension between the Apollonian and Dionysian. Clearly distinctions between optimism and pessimism don't make sense unless we are first clear what matters.

Heidegger is obsessive in methodically, painstakingly approaching his topic—that of Dasein's fundamental attunement. He focuses on boredom! We need to labor to understand how bored we are and its status as a fundamental attunement. This becomes clear when he says that we can be bored without knowing it. Moreover, he leans over backward to distinguish his philosophical account from those circulating in the wider culture, further insisting that he is not talking about feelings or emotions in any interior sense. Attunements are relational, they bind us to others immediately, and they can be shared.

Boredom as Attunement

It is perhaps surprising that Heidegger fastens on boredom as the fundamental attunement of his age. He analyses its superficial form and its deeper ontological significance, and he shows its essential connection to time, even more obvious in the German *Langeweile* (*lang* means "long," *Weile* means "while" as in "a while"). Boredom, specifically, discloses our relation to the world as such, for we are not just bored with this or that. Boredom is an example of a mode of attunement.

This account raises many questions for us here:

1. What (if any) is the connection between boredom and other modes of attunement such as anger, joy and grief?

2. Is Heidegger saying that boredom is the *dominant* mood of his time, or one of many that coexist?

3. Is boredom a cousin of what Marx called alienation, and if so, what differences in sociopolitical perspective does it represent? Heidegger's example—waiting for a train, glancing at one's watch—would look very different on an assembly line undergoing time-and-motion studies. Either way, boredom seems to reflect a kind of temporal alienation.

4. He was giving these lectures at the time of the Wall Street crash (late October/early November 1929), which triggered the Depression, the rise of the Nazi Party, and the election of Hitler (1933). By 1935 Heidegger is no longer talking about boredom but about deep spiritual decline (so deep as to be almost invisible to itself), hatred and suspicion,[9] and then shock, plight, and distress.[10] How are we to think of this change? (At that same time, in 1935, Husserl wrote, "The dream is over.")

5. Can we understand these fundamental attunements as enabling, debilitating, or what? Can philosophy do more than describe them? (Think of Marx: "Philosophy has only interpreted the world. The point is to change it.")

6. What relation is there between Heidegger's discussion of boredom and Nietzsche's sense of the burden (of history)?

Translated into our time, we can begin to answer some of these questions. If I were to imitate Heidegger's references to Spengler, Scheler, and Klages, I would reference books by Naomi Klein (*This Changes Everything*), Michael Bess (*Our Grandchildren Redesigned*), Catherine Keller (*Cloud of the Impossible*), or Thomas Berry (*The Great Work: Our Way into the Future*), and perhaps Jane Bennett's *Vibrant Matter*. They fall into categories such as revolutionary politics, technological optimism, new spirituality, and vital materialism, respectively—all philosophically salient without being strictly "philosophy." Each testifies to a convergence of material and intellectual/ spiritual crisis, and anticipates or performs a radical transformation.

At this point I feel compelled to paint a rough sketch of our time (2016/2017), if only to come clean about my pre-philosophical grasp of things. And with a thousand apologies. I am working with a picture, that I know to be a picture, but I need to share it, for it is otherwise the elephant in the room.

A Rough Portrait

It is hard to deny that this is a time of Anger. ISIS is angry with the West, and with Muslims who won't sign up to the caliphate. Bernie is angry with Wall Street. Trump is angry with illegals, the media, and the establishment (including the Republican establishment). UK Brexiteers are angry with Brussels and the European Union. Blairite British Labour MPs are angry with socialist Corbyn. America was more than angry with those claimed to be behind 9/11. More positively, Trump, Corbyn, even ISIS represent courage, honesty, and hope for their supporters. Not to mention the sweet taste of revenge. But, on the negative side, anger is only the first of a string of appropriate words here. Close behind we find fear, anxiety, insecurity, a sense of betrayal, nostalgia, injustice, disappointment, and disempowerment. Here is Emmanuel Macron on being elected French president in May 2017:

> I understand the anger, the anxiety, the doubt which many of you have expressed and it is my responsibility to hear that. A new page in our long history is opening. I want that page to be one of hope and renewed trust.

And terrorists presumably seek to cause terror. Many of these moods and attitudes and passions could be included within the broader ambit of Nietzsche's *ressentiment*. For Nietzsche *ressentiment* was associated with slave morality, which is essentially reactive. It blames the evil other for the suffering one is undergoing, and is characterized by projection, sometimes quite justified. If the other is seen as powerful, blocking my own agency, *ressentiment* can turn to bitterness. At this point, resignation and despondency can set in, or a fuming discontent open to being channeled by those who seem to offer a way forward. It is not difficult to see how some such scenario opens the path to both revolution and authoritarian dictatorship. Into any explanatory mix we would have to stir some of the language found in but not exclusive to psychoanalysis—such as projection, repression and displacement. Arguably the ethical turn in deconstruction, especially after the Levinasian input and its attention to the pathologies of the Other, has been provoked by awareness of the legacy and continuing threat of just such dark forces.

This somber portrait of the present age would not be complete without reference to the powerful sources of hope, collective organization, courageous resistance, and experimental forms of living. The worry is that isolationist, authoritarian tendencies might have the edge, tapping in to our reactive lizard natures under crisis, panic, and emergency conditions (see 9/11).[11]

If we step back to ask more deeply what is going on, there are clear analogues of the times in which Heidegger is writing. The 1930s in Germany were an economic disaster caused indirectly by the 1929 crash and the burden of previous war reparations, leaving millions desperate and looking for scapegoats. Today global trade (free markets), outsourcing, and migration of labor have led to deindustrialization and middle-class job losses in the West, leaving people fearful of their future prosperity. White American men, their wage earning powers threatened, joined Promise Keepers or the NRA. And they applauded when Trump promised to build a wall that would keep out Mexican thieves and rapists. Many in England joined the right-wing UKIP. For many of those who are unemployed, underemployed, or stuck with low-wage service-sector jobs, times are tough in northern England, the industrial heartland of America, and Greece. It is not difficult to feel cheated by the big banks, the Establishment, Europe, or liberal elites. And if you're a straight-up white guy, after feeling threatened by Islamic extremism, it is easy to feel threatened by black power, women's rights, gays and transgender folk, or a black president. Poverty, poor prospects, and reduced opportunities are felt as attacks on dignity, identity, self-worth. This is perhaps the special source of power of what we might call moral emotions—mistrust, disrespect, betrayal, outrage, indignation, and humiliation. Justified or not, they often reflect a wounded sense of self ("I deserve respect because I'm male, white, English-speaking, handsome," or "I was born here," or "I used to be a famous singer"). Identity investments, very much tied up with stability through time, are notoriously hard to challenge or relinquish.

A Simplified Narrative

My narrative here is an undertheorized simplification, and it may be a media construction. But it contains more than a grain of truth. And if it only roughly captures a minority mood, such groups can change the course of history. Let us suspend judgment on its truth or adequacy for a moment. Heidegger would call it journalism, articulating a cultural meme at best. But what happens if we try to think through the deeper implications of such a mood? What is being registered as malaise, anger, fear, or anxiety about life prospects on the part of white working-class men in the developed world, having to compete with low wage workers abroad, is likely an effect of globalized free-trade driven by the mobility of capital and profits (very much not subject to democratic control). This system is driven by a market economy, aided by high technology and accelerated information flows, which everywhere seeks to maximize profits. This is tied to the refusal to acknowledge the significance of natural capital, the commodification of natural resources, the general project of the domination of nature, and the externalization of ecological and other costs onto such sinks as the oceans and the atmosphere. These market forces have no interest in human welfare, or even in the long-term health of Earth's systems. In this perfect storm, political anger can be easily pressed into the service of defiant and rampant environmental destruction.

   In the first instance, this is meant as an account of "how things work," not as a criticism. Big businesses have the psychological profiles of sociopaths.[12] They have to pursue the quarterly bottom line, given the existing rules, or they go under. That's zooming out, looking at the big picture. We can already anticipate ways in which this mood molecule—*ressentiment* as shorthand—demands a temporal interpretation, as Heidegger insists more generally. Let us now zoom back in. Heidegger is famous for insisting that dwelling is not about housing, care is not about nursing, and we should not confuse the ontic and the ontological. But we must be very careful here. In a world in which most need money to survive and to flourish, and one typically needs a job to have money, unemployment has a direct impact on one's manner of dwelling, even in a deep sense. A job may be neither a sufficient nor a strictly necessary condition for flourishing, but it is commonly so. Why? Because it brings a certain security about the future, which itself means you can put down roots, cultivate friendships, plant a garden, visit the library, listen to music, and walk the dog. Each of these activities can be described in mundane ways, but also in ways that bring out what is at stake in being able to engage with them. In gardening one can bear witness to *physis*, honor nature rather than try to dominate it. In friendship one can cultivate respect for the other. In leisure activity one can step back from everydayness and read Plato.

No doubt economic security can lead to boredom, complacency, or an unreflective life. But it is a romantic fantasy to think of insecurity as the great stimulus to creativity. If we zoom out again, we can see that the shrinking and impoverishment of the middle-classes is part of a historical process in which legitimate expectations will often be dashed, as people and their skills become increasingly substitutable, dispensable, calculable. It will be said that this is not new. Surely the Luddites saw it coming, as early nineteenth-century English textile workers were displaced by looms. And didn't Marx thematize much of this when talking about alienation?[13] The difference, surely, is that the confluence of globalization, information technology, and the limits to growth set by climate change suggest that the old politico-economic order of the West may be coming to an end. Heidegger gets close to saying this: "Yet who speaks [of indifference yawning at us of all things] when world trade, technology and the economy seize hold of man and keep him moving"?[14] If vast numbers of people are no longer needed, surplus to requirements, their expectations of a good life will be disappointed. Democracy could give way to riots, with no obvious solution in the offing.

To put this very quickly, today's mood (anxiety, anger, *ressentiment*, and so forth) is tied both to short-term frustration (which may be cultural as well as economic), long-term anxiety (what are our prospects?), and ultimate unsustainability (perhaps only dimly glimpsed). It is too easy to say that these are all problems of calculative time. Security, predictability (up to a point), and confidence in one's ability to plan ahead enable a range of virtues. Negative capability, as it has been called, is something of a luxury not available to someone who does not know where the next meal is coming from.[15]

If this account of today's mood has any merit, it is especially important to realize that critical reflection on it can open a quite different future. We cannot go on like this.

SEVEN

Posthumanist Responsibility

> There would be no decision, in the strong sense of the word, in ethics, in politics, no decision, and thus no responsibility, without the experience of some undecidability.
>
> —JACQUES DERRIDA, *"Hospitality, Justice, Responsibility"*

> Everything takes place as if I bore the entire responsibility for this war.
>
> —JEAN-PAUL SARTRE, *Being and Nothingness*

The issue of responsibility in connection with global climate change is especially challenging. In some ways, it parallels the logic of the firing squad (all blanks but one)—we are each absolved of responsibility. "I didn't melt that glacier." And yet, very likely "we" did, even though there is no collective "we" that acted.

The more we know about the excessively large typical Western carbon footprint, the more easily we each can feel guilty—about travel, our lifestyle, our food, and so on. This experience cuts through the lack of a collective agent, drawing on some such idea as participation: "I myself didn't melt the glacier, but I represent a lifestyle that did, and I see myself as a prime example of that."

I want to consider some fundamental questions about the nature of responsibility, especially in the face of doubts about the agent-as-subject, from posthumanists, new materialists, feminists, deep ecologists, and others. But before that, I will sketch the landscape of such responsibilities as we may suppose we have, and why these issues are both pressing and puzzling. I will also bracket for the moment the question of who "we" are ("we" who might be responsible).

Angles of Responsibility

There can be no responsibility without a patient—those to whom we are responsible. Various answers thrust themselves forward—our species, other species, the earth, future generations, and last, perhaps, what I call the fading promise of the human. The Brundtland Report's (1987) understanding of sustainable development is a principle of responsibility: "the kind of development that meets the needs of the present without compromising the ability of future generations to meet their own needs." It is officially entitled *Our Common Future,* capturing the sense that this responsibility is, in a sense, to *ourselves* as an ongoing species, and that we have the same responsibility to those temporally distant (in the future), as we do now to those in distant parts of the globe.

It is not difficult to sign up to this, but it does raise troubling questions. Some doubt whether we can adequately anticipate what future humans will need, or think they need. Some worry about whether future humans would still be sufficiently like us for us to care (Imagine future kids as couch-potato stalks attached to screens and hard-wired pleasure receptors.) Many want to know how far into the future we are talking about. Even if they will be very much like us, can we really care beyond the horizon of our own children? Is it a stretch, or a solid plan, to be calculating safe storage for nuclear waste 10,000 years into the future (with universally understandable signs marking the sites for future civilizations)?

Responsibility and Imagination

What these examples show is that there is a deep connection between responsibility and some sort of empathic imagination. If we see our responsibility as extending to the earth, as Gaia, as "our" mother, or home, or to the other species with whom we share this space, something like imaginative connectedness is unavoidable. There are some exceptionally ugly fish that occasionally get brought out of the ocean depths. We can come to care about them by recognizing them as living creatures, like us, and by a kind of transpositive imagination, in which we translate their shapes into ours or *vice versa*. If we identify with the life stream, it is because we can see ourselves, *inter alia*, as part of it. This is not at all to deny that the ethical begins with the Other with whom we may struggle to identify. I am saying that the ground for even that struggle is some recognition of continuity. The ethically significant Other typically does not include rocks, clouds, or puddles.[1]

When we consider other species (and indeed our own), the issue of responsibility is fractured once again by questions of singularity and generality. Am I responsible for your children, or just my own? Am I responsible for specific nonhuman creatures (like my pets) or for all such beings, or (quite distinctly) for the survival of species? These issues are not just old philosophical chestnuts; they are of urgent significance. And what these examples make crystal clear is that questions about "my" responsibility just scratch the surface. For immediately the question of "we" arises, the ways in which responsibility is amalgamated, distributed, and shared among different "we's" some of which may yet need to be constituted. We may have a responsibility to form we's that could effectively discharge our responsibilities. This strange thought may justify the thought that ought need not imply can. It is often said that we can only have obligations that we are in a position to fulfill. But our not being able to fulfill them might itself be something we ought to address. This would take us down the path of forging social networks to effect change that would otherwise be impossible.

These preliminary ruminations have hardly touched on the nature of agency itself. They are largely compatible with the classical idea of an autonomous agent-subject from which an act flows as its origin. Responsibility is *to* the other, and rests on a certain self-relation. It is typically tied to freedom, guilt, and bad conscience.

Is Responsibility a Metaphysical Leftover?

Against this way of thinking of responsibility there is a broad critical counterthrust for which this all reflects an outmoded humanism, harboring a discredited sense of the autonomous subject. Indeed, it is proposed, we should better understand responsibility as response-ability—recognition of vulnerability, openness to the (impossible) event and so on.[2] Nietzsche is a powerful spokesman for this view. He once described free will, as we have seen, as a "hangman's metaphysics," implying that free will is not so much a matter of fact as a legitimating ideological implant. Free will establishes responsibility, which generates guilt (felt or otherwise), and justifies punishment. How helpful is this caution with respect to guilt and responsibility in thinking of "our" situation in an era of global climate change?

The critique of the autonomous subject as the product of implants enabling social control has a point. It is evident that many of what we take to be our desires are ones we have absorbed from manipulative advertising and do not actually bring us the promised happiness. Our purported autonomy would actually be a kind of enslavement to public moralism.

The Subject and the Bathwater

But what of the "rational being" that can diagnose such distortions and set about rectifying matters? And to press the point home, what of the being that recognizes the limits of a certain overly rationalistic model of the subject, and tries to address the problem indirectly—through meditation, participation in alternative community building, or performing music? What, after all, of the writer who carefully dissects the errors of subject autonomy by giving judicious credit to those who have paved the way, and who tries to persuade others of his views? I am saying that we should not throw out the baby with the bathwater. The autonomous subject should not be reduced to being merely a metaphysical nose-ring by which we are led around. The critique of consumerist conformity only works given some alternative sense of independence, even if one is equally unhappy with the language of authenticity (see Adorno on Heidegger[3]).

Moreover, an ideologically reductive repudiation of subjective autonomy could itself just as plausibly be seen as an alibi-generating device for those who genuinely are responsible. And here the question gets political. Suppose for a moment that we accept that fossil-fuel companies, and their climate change–denying smokescreen minions, are more than passive enablers of current levels of CO^2 emissions, but perhaps, "responsible" for them. In saying this we would not be committing a metaphysical blunder, reverting to some discredited humanism. Instead, we would be asking: Should we hold to account those "responsible" for decisions about drilling safety on oil rigs, toxic spillages, funding anti-climate change think-tanks, and promoting the burning of fossil fuels at every turn? And if these CEOs quote their fiduciary responsibilities as corporate executives to their shareholders and to the quarterly bottom line, should we not be talking about changes to the laws of incorporation? Worse, if they hired a posthumanist lawyer who spoke of shared diffracted hybrid responsibility, would it not sound like a scarily inverted reprise of Eichmann's Nuremberg appeal to Kant and "following orders"?

The movie *The Corporation* (2003) presented a case for corporations having the diagnostic profile of psychopaths—disregard for the feelings of others (or their safety), absence of guilt, disrespect for social norms and the law. It is, of course, ironic that corporations are legal "persons" under US law, at least in terms of their freedom of expression. My point is that we need the language of accountability to stick around at least long enough to imagine ways of legally restraining these antisocial traits. It would not be ironic but tragic if posthumanism were to provide solace and cover for corporate psychopathy.

There is however a major limitation to the focus on straightforward agency that would, if addressed, actually increase both guilt and responsibility. I am talking about negligence. This got a lot of press in the case of the Deepwater Horizon oil spill (2010), and it can be generalized.

Culpable Negligence

In this second take on responsibility, I focus not on what one does but on what one fails to do.[4] The strong legal version of this is culpable negligence; arguably much environmental damage is the consequence of what could and should have been avoided. The world in which doomsday is the result of evil men selfishly plotting the earth's destruction is a Mickey Mouse world, a cartoonlike simplification of a complex reality. And yet it would be equally foolish to pretend that there are not powerful people making strategic decisions that have predictable destructive real world environmental consequences. When Chomksy claims that the (US) Republican Party is the "most dangerous organization in human history," as he did in May 2017, this is what he means.

Once deposed, dictators of small African countries now know that they may not escape the attention of the International Criminal Court, if they are party to genocide. It's not a good time to be President Omar Al-Bashir of Sudan. It would be a real step forward if executives of fossil fuel and other carbon-unfriendly major corporations were to feel the same uncertainty over their decisions. I am not just talking about flagrant lack of concern with safety (e.g., oil spills), but about their broader role in sustaining fossil-fuel addiction.[5] If this is not currently on the cards, it is partly to do with "plausible deniability."[6] If catastrophic climate change is still a matter of debate, it is plausible to argue that there was no truth that one was failing to acknowledge. This is surely one of the motives behind the funding of right wing think tanks in their climate denial, which has been compared to Holocaust denial.[78]

Pursuing the line of culpable negligence would apply to our political representatives, as well as executives. It too complicates autonomous agency, with respect to both autonomy and agency. Leaders and executives do have special responsibilities, and it is important that they be held to account. Their position and power rests on their projecting and working toward distinct visions of the future in concert with the goals and aspirations of the institutions with which they work, and those to whom they are answerable. And a version of the response to Eichmann's defense at Nuremburg needs to be made to them too. It could be a criminal activity to knowingly lead people to a disastrous fate for personal or corporate gain, like, say, a captain on a North African immigrant vessel.

It will rightly be said that these are the easy cases—people in positions of power recklessly failing to act. Are gun manufacturers responsible for crazy people going on a rampage? Are fertilizer manufacturers responsible for the nitrogenous runoff that causes algal blooms? Pharmaceutical companies for the misuse of their meds? Exxon-Mobil for climate change (or just one of many players)?

But there is an argument for extending the law on strict liability. Strict liability applies, for example, to dynamite, and owning wild animals, and it does not require intent (or even negligence). Companies manufacturing dangerous products could be made strictly liable, not least to give them an incentive to monitor their safe use.[9] This would make contributing substantially to dangerous negligence quite as culpable as sole responsibility.

For all this to be possible it is vital that the public sphere not be held hostage by climate-change deniers. If we needed it, this gives all those who aren't in the pocket of the oil companies and their satellites a particular responsibility to speak up, to speak truth to power: parrhesia.[10] We need a tipping point in public opinion.[11] (Only 28 percent of Americans think climate change should be a top priority.[12]) Downright lies corrupt the public sphere.

Catastrophic climate change would likely be a nonlinear tipping point event that we cannot *know* will happen ahead of time. But the possibility of it happening and that its happening (or not) might well hinge on small things makes activism at every level worthwhile. There is a clear convergence here with Heidegger's reference to "preparing the way" when writing about an Other Beginning. We cannot just prevent disaster singly or collectively by a resolute decision before breakfast. But we can work to change the background conditions—especially public opinion—that would help resist it.

I have been defending a broadening and deepening of a fairly traditional sense of responsibility, one that steps back from the autonomous subject in ways different from posthumanist critique. This expansion does not shy away from attributing "guilt" and "responsibility" in the same spirit that Derrida concedes the vital if strategic importance of fighting for human rights, even as he might debunk the idea theoretically. It would reinforce the idea that willful or culpable ignorance is no defense.

It is worth adding here that guilt may sometimes be stimulating and productive, and not just a source of depressing inertia. Visiting Australia not long ago I reflected on the fact that flying to Sydney from Nashville releases the equivalent of 12.3 tons of CO^2 (including nitrous oxide, particulates, and other greenhouse gases). This is something like two-thirds of the average American's annual carbon emissions. I could instead switch to a gas-guzzling SUV, chop down a big stand of trees, go back to eating steak and still come out ahead. This is, I concede, a source of concern, both practically and morally. Is it not just such a refusal to alter our habits that is killing us? And can I really think that I can sweet-talk my way out of this? Surely we should be meeting on Skype, and forget about pressing the flesh, eating and schmoozing together? Am I paying off my guilty conscience by this very confession?

Response-ability

But there is a third way of rethinking responsibility that deserves attention. Instead of thinking of it as the internalized property of an autonomous subject, we can think of it (broadly following Heidegger, Levinas, and Derrida) as a response-ability. It becomes a kind of openness to the other, one that cannot be exhausted by rule-following, one that obeys no algorithm, and one which inevitably leads to a gap between one's responsibility and one's ability to discharge it. This hyperbolic responsibility reappears as an openness to the impossible (to what I cannot myself will or predict), which exceeds the space of my self-understanding. This maps quite well onto my justification for small-scale environmental activism—that it just might be the difference that makes a difference. Tipping points are not under our control but we can position ourselves better for them. The impossible may in fact be imaginable, even if how we get there may not be. What Abraham would call his faith in God we might call openness to the impossible.[13]

In his *Contributions*, Heidegger steps back definitively from the will, from the willful pursuit of one's objectives (with Nietzsche and the Nazis in mind) and speaks instead of preparing the way.[14] He is alluding in part to Hölderlin and his soliciting the return of the gods. What is of interest is that there is a certain agency at work here, but it is indirect. And in terms of overcoming pessimism, galvanizing motivation, this formulation is very helpful. We can take the horse to water, but we cannot make it drink. We can steer things as best we can in the right direction but we cannot force the outcome. The fantasy of fomenting revolution was that one could create the conditions in which an exploited class would come to see itself, for the first time, as a "we," a subject. (Chapter 1 above, *Herding the Cats of Deep Time*, encourages that with respect to our species-being.) This would indeed be a tipping point. But unlike the narratives of Hegel and Marx, there's no necessity being invoked by a Derrida, and there is no specifiable hero of deliverance. The im-possible may turn up out of left field.

Derrida is not unaware of the problem that opens up here. If hospitality (to the future) is another word for responsibility, he will write that true hospitality (if such a thing exists) would admit even the stranger who would burn down the house.[15] What are we to make of this? Do we learn what we need to learn by going hyperbolic in this way? One thing is clear: Going hyperbolic makes sense only if we abandon the agent-subject called on to act, or abandon the assumption that ought implies can.

In the next chapter I give qualified support to the effort to develop hybrid forms of agency proposed by some proponents of new materialism. But before exploring their agenda, I want to give voice to a metaphysical consideration that should not go unaddressed, for it importantly sets limits to our avowed realism (for which the universe just is some 4.5 billion years old), and it gives some insight into the attraction of the idea of the agent as subject, even as that is a flawed ground for geologically responsible resistance to climate change.

Geological Freedom?

Crudely put, why does the geological perspective not cast a materialist shadow on *any* idea of "responsibility," "agency," "progress," even "consciousness" itself? The old chestnut is this: how can there be freedom in a mechanical universe? We imagine a little play in the steering wheel or atoms that swerve (Lucretius).[16] But this basically rejects the mechanical premise. We have been gripped by the model of the clock—a seemingly closed system in which interconnected parts frictionlessly click forward. We imagine that "external" factors like temperature, humidity, barometric pressure, and the like can be calculated precisely. But this is an article of faith. The laws of science can only be demonstrated under controlled conditions, and when the results can be replicated. As such, science is inapplicable to any singular historical event, for these can never be exactly repeated. The French Revolution, the Vietnam War, and 9/11 *could not*, in principle, be repeated. And while this may be obvious in these momentous instances, it is surely possible that this is true of any event whose constituent variables are exposed to the world. If science demands precise repeatability, then its necessities and imperatives may be limited to the empty set of absolutely closed systems.

Following this line of argument, there is no need to make room for "freedom" because determinacy in the real world has not been demonstrated. It is little more than science fiction. Another line of defense would be to argue that this whole nomothetic view of science is hopelessly limited. There are "field studies"—most notably ecology itself, where controlled conditions do not obtain, and there is a premium on careful observation and cautious induction. This does not seem to require or support determinism.

Suppose however that there is some unsuspected flaw in these arguments. Suppose a fully blown materialism is ultimately persuasive, that lab science is the gold standard, and we somehow concede that the findings of controlled science experiments can be extended to the real world. What then of freedom, creativity, and real agency?

Irreducible Positionality

Materialism is a theory, a position, an account of things; to forget this is a fatal naivety. Minimally it requires that at least some of the items we call matter (like humans) are complex enough to hold theories. But this attempt at including point of view in a view from nowhere is bound to fail. This is the logic of Husserl's *Logical Investigations*, Kierkegaard's *Concluding Unscientific Postscript*, and Nagel's *View from Nowhere*. It fails not because the content of subjectivity cannot in some sense be taken account of. It may well be that multiple perspectives can be pulled together to form an image of the whole like piecing together the fragments of a broken figurine. If subjectivity means "partial perspective," the partiality aspect, as content, may be susceptible to correction. It might then be argued that the standpoint of the whole is still a standpoint, that "objectivity" is a judgment or a value, or a designation presupposing a subject making such a judgment. To which it may be replied that any subject making such a judgment means to point beyond that gesture to the real "objective" content.

So, what is the problem? There is a more decisive consideration. One thing we know about the world is that it includes what we will call positionality as an irreducible feature. This itself can be grasped objectively or subjectively. From a plane, we can look down on cars on the road or people milling in the square, and after initial comparisons with ants have worn off, we realize that we are seeing a plethora of perspectives and desires. And it is easy to suppose, as we look down on them, that that plethora can be superseded by a new synthesizing vision, perhaps mine. But, as Kierkegaard has shown, the truth is not in the whole. It is a vision necessarily blind to the positions it integrates. Perhaps "blind" is a revealing expression. "Sight" sets up everything up as a matter of coordinated perspective, with the promise of a conspectus, "blind" to any principled resistance to that logic.

What then is meant by irreducible positionality? It is not just that the world includes a bunch of perspectives. But that the very idea of the (whole) world is undermined by this phenomenon. We seem to be able to imagine an anonymous universe in which there is just inert matter. We know, however, both that it consists of multiple centers of significance, and that we are each such. And that there is no reason or ground to account for how or why you or I are *this* one, rather than *that*, or not at all. I call this irreducible but of course it's not *actually* irreducible. Perhaps we should say—irreducible without violence.

Navigating what follows from this is tricky. At first blush it may seem like a hysterical plea for attention—don't forget me!—the feeling a child may have in the grown-up world, or a stranger in a foreign land. Nothing could be further from the truth. Positionality is not about "me," not about "this" me, but about "this-me"-ness. As Heidegger put it, "Dasein is in each case mine." In an older vernacular, this is a transcendental not an empirical claim, though it has its empirical (ontic) correlate. But what is the significance of this claim?—that every human (at least) is "this someone." It is not (at first) the claim that I am not you, but that "you" and "I" (and "him" and "her") are all this-someones. Recognition, conflict, desire, will come later. But first there is the "miracle" of "this, here, now, me," whether or not this is grasped as a miracle (something quite out of the natural order). This does not mean that I am special in comparison with others. But that I am not a rock, not a cloud, and not someone else. Nor does it mean that I must realize this is true of you too, and indeed our practical recognition of the universality of this singularity may not be consistently realized. Reflection, however, makes this step irresistible. For both Hegel and Levinas, it is the confrontation with the Other that brings home the truth of my condition. In Levinas's case, that "truth" is ethical, indeed the opening of ethics. Part of what the face-to-face relation means is "Don't kill me!" But it is at least arguable that the urgency or significance of this plea rests on my first bearing witness to my own desire to live. Perhaps I should say "this" desire to live, because I need not think or say "my." The thought is impersonally personal.

This may be why reflection has such an easy time, because it is a small step from "this" to "any this," compared to the step from "I" to "Thou" or "you," of "the Other." "This" has no specific content. The claim is not that you (for example) are like me in some respect (educated, white, Western). Consideration of nonhumans, and penumbral humans (brain-dead, infants) raises important issues of mode of this-ness, not of "content."

This haecceity[17] interrupts materialism, but not with a clinamen or an in-determinacy, and not with a different kind of substance (spirit). Although it is tempting to suppose that this was the point of references to Spirit in the tradition all along. (See Wittgenstein: "Philosophy leaves everything as it is.") Nothing has changed and yet everything has changed. It seems there is one thing to say here, but there are two. With Descartes or Kant we can point to self-reflection, one's awareness that one exists, the "I think." But there is also the concomitant and less visible awareness that "I am *this* self-consciousness." Self-consciousness is a manner or quality of existence. That I am this one is in some sense unthinkable. It corresponds to no concept, and no imaginable concept. It perhaps nudges up against Heidegger's being toward death, grasping the possibility of my impossibility. It is further captured I believe by Derrida's strange claim that each death is the end of the whole world.[18]

Whether we link ethics to freedom (Kant, Sartre, Spinoza), dwelling (Heidegger), or to radical passivity (Levinas), what they share is an under-determination of the content. Indeed, much of the sense of responsibility that each involves has to do with the work involved in figuring out what to do, and the sense of deepening possibilities of centeredness. Where does this responsibility come from? For Levinas it is the challenge presented by the Other in need (widow, orphan, stranger)—the challenge to justify "my place in the sun," as if my being alive is at the expense of others, a zero-sum game.

Being Alive

But it is not obvious that this interruption can even happen unless one has already developed a sense of what it might be to be alive—contingently, unnecessarily, and for such a "short time." For Hobbes, it was life outside society that was "solitary, poor, nasty, brutish, and short." Need life be "a tale, told by an idiot, full of sound and fury, signifying nothing" (*Macbeth*)? Do we not need a certain ontological work before arriving at such conclusions? We can indeed try to understand (or come to terms with) the specific "mode" of one's own (and indeed anyone's) existence. It is perhaps too harsh to say that "the unreflective life is not worth living," but one's life is surely a compelling subject of reflection. Its pursuit is very far from narcissism, both because it can have powerful and positive consequences for others, and because the existence and welfare of others can be expected to play a major role in self-understanding. So, if an ethical opening is provided by the "miracle" of thisness, it lies in the challenge of thinking that it provokes, to try to figure out what it might mean. The ethical dimension arises precisely because it opens up a sense of one's deeper connectedness to others (and the earth) at the heart of one's subjectivity.

We earlier set aside the argument that the application of science to the real historical world is limited because the evidence for its truth rests on perfectly replicable experiments, which are not only themselves strictly speaking not possible, but certainly never found outside the lab. Practically speaking, it will be said, we do believe in science. And no doubt we are right to do so. Repeated takeoffs and landings are sufficiently reliable to induce trust in the surface implausibility of air travel. But such practical assurance is no basis for the strong metaphysical claims that some are then tempted to make. We do not and could not know what normally recessive or absent variables might affect the applicability of the general scientific truths that, *ceteris paribus*, serve us well. The importance of such complexity is clear enough in medicine. The fact that there are many different, often overlapping natural sciences, no clear prospect of a unified science, and respectable modes of scientific inquiry (field studies and research) that do not conform to the covering law model, further widens the metaphysical gap. (Should we turn to physics or biology or ethology to understand an ant?) We set aside this argument because, however convincingly it may limit our faith in science, it does not specifically address the issue of whether "positionality" is a fundamentally disruptive dimension. I am not thinking here about questions drawn from relativity theory, but rather whether human agency is threatened by a geologically informed materialism, especially one in which we find ourselves largely observers of planetary destruction, driven by forces we do not control, rather than agents.

I have argued that positionality is ineliminable and irreducible. Science may not "think," as Heidegger claims. But it is nonetheless a way of knowing, and hence oddly presupposes positionality. It is a mistake to try to *picture* how agency escapes causality (the problem with the swerve). Rather (following and generalizing Kant's lead) we have discourses of agency that interrupt those of Olympian passivity. The problems of geologically informed agency in the face of global climate change are not themselves metaphysical but social and institutional.

EIGHT

The New Materialism

> If at the foundation of all there lay only a wildly seething power which
> writhing with obscure passions . . . what then would life be but despair?
>
> —SØREN KIERKEGAARD, *Fear and Trembling*

> [From] a properly cosmic viewpoint, our entire planet (together with its
> flows) would itself be a mere provisional hardening in the vast flows of
> plasma which permeate the universe.
>
> —MANUEL DELANDA, *The Geology of Morals*

Posthumanism throws up deep questions about agency and responsibility
at a point in human and terrestrial history at which we most urgently need
answers.

I want here to tap into an unlikely source of relief, the New Materialists,
a cluster of contemporary thinkers influenced by Spinoza, Deleuze, and fem-
inism.[1] Is it enough to speak of the agency of things? Or a hybrid model in
which humans share agency with natural forces? Or to rethink agents not
as independent beings but essentially relational assemblages? Does new
materialism fare any better than the old ones? Let me explain first why I
call it an unlikely source of relief.

Thinking geologically, understanding human beings on a time scale
much greater than our own history (or even the history of life), invites a certain naturalism, in which our pretensions to being special are burst like a
bubble. The old materialism had little space for such values as responsibility.
Nietzsche began his famous essay on Truth like this:

Once upon a time, in some out of the way corner of that universe . . . there was a star upon which clever beasts invented knowing. . . . After nature had drawn a few breaths, the star cooled and congealed, and the clever beasts had to die.[2]

According to Nietzsche, the cosmos will not have been affected by this brief human blip. "When it is all over with the human intellect, nothing will have happened." Our sense of self-importance, of being the flying center of the universe, is shared by every gnat buzzing in the air. This is a deflationary take on human exceptionalism, one that Foucault echoes when he compares the idea of Man to a mark on the beach erased by the tide.

Obviously when we speak of *thinking* geologically, we are taking the thinking part seriously, not just the geology. But how does thinking escape being seen itself simply as a natural phenomenon on a par with breathing, or indeed with raining, or congealing?

A Kierkegaardian Response

The classic nineteenth-century "response" to Nietzsche can be found in Kierkegaard's *Fear and Trembling*. To expand our epigraph,

> If there were no eternal consciousness in a man, if at the foundation of all there lay only a wildly seething power which writhing with obscure passions produced everything that is great and everything that is insignificant, if a bottomless void never satiated lay hidden beneath all—what then would life be but despair?[3]

Kierkegaard claims such a view is unthinkable, not to say deeply dispiriting. But analogs of it, and Nietzsche's Olympian distance, have reappeared more recently. Here, for example, in a Deleuzian spirit, is Manuel DeLanda:

> Our individual bodies and minds are mere coagulations or decelerations in the flows of biomass, genes, memes and norms. . . . Given long enough time scales, our languages are also momentary slowings-down or thickenings in a flow that can give rise to a multitude of different structures.[4]

Performative Self-Contradiction

But is there not a massive, pervasive performative self-contradiction here? Isn't that what Kierkegaard would say? There is a claim to truth, even as truth is being dissolved in flow. How can this be so? DeLanda writes (see the epigraph) "from a properly cosmic viewpoint"—but what could that mean? Surely, unless you are invoking God, the cosmos does not have a viewpoint. Where would it be located? I struggle with DeLanda's cosmological formulations. He wants both to reduce minds to brains and brains to self-organizing flows. The idea that we might productively live on the edge of chaos, while preserving a safe haven to return to, seems awfully romantic. But it does capture the sense that we may need our traditional organizing categories for a certain stability, even as we experiment with new languages and formulations.

The same problem seems to rear its head in the work of Timothy Morton.[5] *The Ecological Thought, Ecology Without Nature,* and *Hyperobjects* each argue, somewhat hyperbolically, for the redundancy of a whole range of concepts (world, nature, place, metalanguage). The advent of the hyperobject (especially that of global warming) changes everything. (Hyperobjects are "things that are massively distributed in time and space relative to humans" [*Hyperobjects,* 1]. Background becomes foreground, and our thinking is turned upside down. Morton is exquisitely aware of the theoretical and practical dilemmas with which global warming presents us, and is a master of interweaving concrete detail with original frame-changing interpretation. And yet he looks to be repeatedly caught up in the problems he addresses. For example he diagnoses the folly, often arrogance, of the meta-stance, in which I can see through you, or I can understand the broader picture in which you are myopically mired. "Going meta has been the intellectual gesture par excellence for two centuries" (*Hyperobjects,* 155). He continues: "This attitude is directly responsible for the ecological emergency." This is either deeply tongue in cheek or an obvious repetition of the very phenomenon he is critiquing. He could respond that this entanglement is exactly what he is talking about, that it is unavoidable, and that it is not so much performative contradiction as performative confirmation. He might add that his distinctive style of writing—personal, playful, conversational—exempts him from these kinds of concern. This is surely the point of Nietzsche's "My style is a dance." And although he does not position himself as an inheritor of Derrida's deconstructive motifs, Morton seems adept at conjuring a version of Derrida's double strategy—immanent critique and the step beyond. In this way he arguably escapes the immediate difficulties we had with DeLanda.

The unresolved question that runs through much new materialism is this: how can we philosophers reconcile our capacity to think about (reflect on or own) our geological heritage and commitment to its continuation, with our growing awareness that we are destroying the sustaining background conditions that make either possible, and the traditional language in which we used to be able to think about these matters? There is not a rift between Man and Nature, but a frayed and aporetic line. The standoff between Nietzsche's disdain for human self-importance and Kierkegaard's sense of ineliminable spirituality, nicely exemplifies the rift. But their own deep grasp of the importance of style, strategy, even paradox shows that we cannot be satisfied with any simple representation of that divide.

Responsibility Facing Two Ways

In talking about responsibility, we have highlighted our responsibility to the future (for example to future generations). But we have responsibilities to the past as well. The most obvious examples would be to those who died in battle, in resisting oppression or in battling for civil rights. "Never again," we say of the Holocaust. And we honor those who gave their lives in war. But as inheritors of a human cultural legacy, we may well feel something like a responsibility to inherit it well—creatively, productively—even as we transform it to our own circumstances. As we saw, discussing Nietzsche in Chapter 3, this will mean both critically honoring the (monumental) achievements of the past, and trying everything we know to avoid its tragedy and disaster.

Whatever we think now about the Enlightenment, which undoubtedly has blood on its hands, we cannot give up on some of the values history has bequeathed to us—truth, justice, compassion—despite the ways in which they can so easily be distorted. I would gloss compassion as something like respect for vulnerability, willingness to be affected, which returns us to what we have called response-ability. Derrida's democracy-to-come is something of a testimony to this responsibility to preserve and project, without always knowing ahead of time what it will look like.

New Ways of Talking

Inheritance, the subject of Derrida's *Specters of Marx*, is where the plot thickens. The twentieth century witnessed a relentless assault on traditional metaphysical thinking, especially on stale oppositions that hinder thought and often function to maintain oppressive relationships. This happened with Heidegger's account of the destruction of the history of ontology (1926), Derrida's deconstruction, Deleuze's transcendental materialism, and Foucault's antihumanism. We witnessed the death of Man, the death of the Subject, the End of Nature, the End of Art, and so on. Classic cornerstones of Western reason crumbled, including reason itself. The plot thickens for us because included in this list is Man as opposed to Nature (or Spirit v. Matter). And it is just such oppositions that Nietzsche and Kierkegaard both rely on in the passages we quoted. If responsibility to the past includes (or perhaps requires) inheriting it transformatively, might we therefore need to jettison the very language that would speak innocently of our agency in relation to nature in man's response to climate change? Might not responsibility first require a whole new language?

This surely is the claim of New Materialists, and of Timothy Morton in spades. They offer ways of thinking and talking that attempt to overcome or displace patterns of thought that would generate blind repetitions of empty formulae. Following Deleuze, they would attempt to overcome the ways in which the human is caught up in reactive forces, which has a direct impact on how we respond to climate change.[6]

It is no simple task to set aside our categorical habits. The nature we are up against is one we are part of and/or responsible for. We have already alluded to the figure of the Moebius strip, a simple band that is cut and then reattached after a single twist. This topological figure elegantly captures what it is so hard to think. At every point there are two sides, but these two sides form a single continuous surface. By analogy, we may construct a situation as one in which a Subject (Man) relates to an Object (Nature), and in which these two seem "opposed," just as there are two sides of the paper. But these two opposites nonetheless occupy the same plane, and are not radically different substances. The difference is at every point something to be negotiated. The Moebius strip captures Morton's refusal of meta-language quite nicely. But its lesson is that even if there is no absolute metalanguage, there are countless circumstantial opportunities for one discourse to speak about another. There may be no canonical Aboutness, but local aboutness abounds.[7]

Sampling New Materialism

I would like to offer a brief but critical appreciation of how New Materialists are currently contributing to rethinking agency and responsibility/response-ability.[8] I will unapologetically sample their work rather than claim to do full justice to it.

The first key idea is that of material agency, which insists that there is agency beyond the human—not just in other creatures but in material forces and other assemblages. As I was writing this, a violent storm blew up, lighting hit a giant tree in my garden. It fell on the new house being constructed next door, catching fire. Without attributing a *telos* to the tree, new materialists might extend to the lightning/tree/gravity complex a certain organized force or agency. But in a way this is too good an example. In her *Vibrant Matter*, Jane Bennett takes up Spinoza's sense that there is a *conatus* (a power) in everything, beginning her list with a dead rat, oak pollen, a stick of wood, a plastic glove, and a bottle cap. Her own more developed example—an electrical blackout affecting a whole grid in 2003 in New York City—is described as a human/nonhuman assemblage, a federation of actants. The grid is a human construction that took on "a life of its own." Additionally, following Merleau-Ponty, Diane Coole speaks of a spectrum of agentic capacities, some located in individual bodies, others in transpersonal processes. Karen Barad, explicating what she calls "agential realism," goes further: "'Agency' is not held, it is not a property of persons or things; rather agency is an enactment, a matter of possibilities for reconfiguring entanglements."[9]

Agency as assemblages or entanglements opens a second key idea. This entanglement is what I call constitutive relationality. Agents are essentially relational (and only artificially congealed into distinct entities), but this re-lation is part of what it means to be something. Barad explains that the "social" and the "natural" emerge through "intra-action" of subjects and objects. Agency is glossed as a "response-ability . . . possibilities of mutual response . . . attuned to power imbalances . . . possibilities of worldly re-configurings."[10]

Jane Bennett and Karen Barad offer tantalizing workshops in how to de-velop the new languages by which a certain sovereignty of agency might be set aside in thinking about our role in the anthropocene. But it is with Nancy Tuana that we find direct and explicit suggestions of what this might look like. She writes of "the significance of embracing the symbolic meanings of the Anthropocene as a vehicle for transforming our imaginary."[11] The figure of the Anthropocene "can serve to disrupt . . . our inherited episteme that sets down a series of imposed subject-positions and dichotomies." This changes what we value, how we live together and opens up "the space of the ethical." As a Whiteheadian, she stresses the inherent interconnectivity among things and the need to shift from a substance ontology to one of be-coming. We must learn how to think and act anew, developing the hybrid concepts of socionature and viscous porosity by which exchanges of mate-rial agency are effected. The human and the geophysical are to be thought together. We need to learn to embrace multiple interrelational causalities, as well as think this complexity in ways liberated from the old oppositions.[12]

The Promise and Limits of New Materialism

This brief incursion into new materialism gives some idea of its promise. That is not to say that it is unproblematic. To my mind, it is more signifi-cant as discursive innovation than it may itself like to suppose. It opens the world in new ways, refiguring matter with new words, new concepts. It wants to point to the resistance, recalcitrance, and powers of things, but much of the work takes place in transforming our discursive inheritance. This is not so much an objection as a correction to its own self-understanding.

I have a second reservation. If we take seriously the increasingly preva-lent mainstream analysis that it is not so much "humans" that are respon-sible for climate change, as unbridled free-market capitalism,[13] then we have to be careful not to allow concepts like the socionatural to obscure this central fact. Too much defracted pluralism and agential hybridity could deflect attention from a deep structural causality. Where new materialism really comes into its own is in helping us develop creative nonreactive responses to complex situations in which whatever the deeper driving forces, a whole range of other factors are in play.

This is a taste of the new materialism and how it might facilitate a geological ethic. The traditional problem with reductive naturalism is that it dampens or excludes freedom and creativity, but new materialism solves this by challenging our images of mechanistic nature and contesting human transcendence. By avoiding the traditional schemas, new materialism avoids the problems they generate and provides an alternative discourse. We can then feed back a new ontology—a new imaginary, a much richer understanding of collaborative agency—into our socionatural practices. If the climate crisis is at least partially a product of the translation of an imaginary of human sovereignty and domination over nature into a free market economy allergic to ecological constraint, new materialism's novel way forward should work at this level. It is not just a new vision (say, of harmony), nor just a new attitude. What this new materialism offers are new discursive practices, new ways of analyzing problems and situations, and new ways of understanding the shape of our engagement.[14]

The (Absent) Elephant in the Room: A Note on Deleuze

Surely this chapter lacks something important—a serious discussion of Deleuze and/or Guattari (frequently referred to as, and hereafter, D&G), who after all, wrote of a "new earth" and "new peoples." Isn't that just what we are arguing for? What could be more promising? How could we give such prominence and not mention Deleuze's stunning *Nietzsche and Philosophy*? And if we are going to focus on the affective, what better guide could there be than D&G's treatment of desire? Have I not taken issue with the smaller fish instead of confronting the philosophical sophistication of Deleuze, their common inspiration? And should I not address the specific attention given to environmental issues, geophilosophy, and chaosmosis by Guattari himself, Gasché, Protevi, Colebrook, Braidotti, and so on?[15]

Although Deleuze himself sees no point in responding to objections, I still see some merit in it. The most straightforward reason to take D&G as allies in the struggle against catastrophic climate change is that they are materialists (albeit transcendental), that they mistrust any attempt to give abstractions that swirl around us explanatory power, and that they have a widespread suspicion of the state and similar institutions. Moreover, they have highly articulated explanations of how desire can be channeled and appropriated to reactionary ends, even when it starts out with deterritorializing ambitions. My gestures toward new forms of community, new shapes of desire, and new kinds of identity investment run parallel to Deleuze's, so why not at least enlist him as an ally?

Having served one philosophical apprenticeship (with Derrida), I am reluctant to inhabit another technical language. As with Whitehead's vocabulary, there is complex trade-off between the power of new concepts and finding oneself unable to connect with those who do not understand them. Moreover D&G's admonition to create new concepts is somewhat in tension with canonizing those they came up with decades ago. Gasché is not alone in wanting to translate them back. This tension is manifestly in play when reading the best of those they have influenced.[16] It is a real tension because of their justified concern that without new concepts we will simply repeat old configurations. And who could underestimate our tendency to backsliding, to reaffirming by the back door the value of identity! However, something of their sensibility is increasingly part of critical culture, even when they are not cited. I hope that is true in my writing here.

D&G endorse utopian thinking, with reservations, partly for spatio-etymological reasons.[17] They often write of a new earth, and new peoples: "We lack creation. *We lack resistance to the present.* The creation of concepts in itself calls for a future form, for a new earth and people that do not yet exist."[18] While this sounds as if it is on board with environmental activism, with "saving the earth," that is not obviously so. Gasché makes it clear[19] that their thinking about "earth," for example in "Geophilosophy,"[20] is inseparable from reflection on what it is for the Greeks to have instituted philosophy itself, tied to the "reterritorializing" founding of cities of free men. When we critique consumerist desire, insisting that, in the face of climate disaster, "we cannot go on like this," and when we put the "we" repeatedly in question (on class, race, gender, species lines), many of the same concerns (and terms) are in play, but it is not clear that we mean the same thing at all by "earth."

D&G demand a complete break with the economy of loss, mourning, which looks back to, or imagines, an identity that we no longer enjoy—personally, culturally, nationally, and seeks to restore it. In its place, so to speak: affirmation, inventing (new concepts, new practices, new relations, new identities), imagining otherwise. It is hard not to be attracted by this joyful vision. But the radical contrast between these two economies is itself something of an illusion. Vestiges of traditional identity may be required to muster the strength to differ, to flee, to betray, to resist. What is the body capable of? Indeed! And we may ask the same of planet earth. D&G write of deterritorialization first applying to the earth, turning away from "the fixed powers which try to hold us back, the established powers of the earth."[21] Here I confess unease. Understood as a description of the impact of technology and industrial production on the planet, this is indeed what happened. But as encouragement—that we should step away from what is "fixed" and "established"—all too easily turns into an endorsement of the techno-hubris that threatens the planet. It is folly to push things to the limit to find out what the earth is capable of. Accepting new identities, new relationalities, new attitudes to death, and new shapes of community is compatible with what we might otherwise call disaster. The economy of loss and reparation is often regressive and reactionary, and we may operate with too narrow a sense of the balance of nature. "Going forward" we need all the creativity we can muster. But to be blind to loss in the face of the sixth great extinction, or climate meltdown, would be to fail to grasp the complex conditions under which life on earth evolved, the interconnected web of life. Do not test what the earth is capable of; it could happily do without us. As a corollary of my doubts here, I must confess my adherence to what Lyotard called grand narrative, which D&G equally distrust. The story that links planetary disaster to capitalism and free trade is too true to be treated as just another narrative.[22]

Following on from this, I worry about another issue often addressed, with mixed success, by Deleuzian theorists. Put bluntly, it is that the destabilization of fixed identities, the creation of new desires, are exactly what consumer capitalism demands. We may be able to explain how some of these will strengthen resistance, while others just induce compliance. But it is not clear that these caveats and safeguards do any actual work on the ground. Am I just sneaking identity back in? No. For it never quite went away, except as a theorist's dream. One cannot maximize decentering relations, multiply oneself as an assemblage, or resist the forces of oppression, without someone somehow putting it all together. Identity does not require essentialism. And, to return to consumer capitalism, while we may embrace multiplicity, employers are pushing its economic counterpart: the gig economy. A connected concern: D&G often give voice to what is called a progressive agenda: Philosophy and art "have resistance in common—their resistance to death, to servitude, to the intolerable to shame, and to the present."[23] "The thinker becomes Indian, and never stops becoming so. . . . We think and write for animals themselves. We become animal so that the animal also becomes something else."[24] But how firm is the ground for this agenda? There are cryptofascist Deleuzians who seem to be wired quite differently.[25]

Finally, as I claimed above with respect to new materialists, I do want to acknowledge the terrain-shifting power not just of D&G's own writing, but also that of his most eloquent interpreters. For while they themselves may not invent new concepts on every page, they do exemplify an inspiring connectivity in writing. Consider Rosi Braidotti, addressing the question of the subject: Deleuze "replaces the old subject formation with a notion of the subject as a cluster of complex and intensive forces—intensive assemblages which connect and inter-relate with others in a variety of ways."[26] Or Claire Colebrook: "it is always a question of *whose* life. . . . If we can do philosophy and criticism without an image of life—if we can free thought from its home in the human sensory-motor apparatus—then we might start to think the difficult and ethical questions. Not: is this or is this not life? But what style of life? How much difference from organized harmonious life—how much thinking can we bear?" The first remark makes it clear that we are not abandoning the subject, but rethinking and rewriting it. The second poses the question we just asked about the earth. In relation to the individual human subject, it is an encouragement, giving courage. But with no formulaic answer. My doubts center on extending this experimental approach to the planet.

If Deleuze is the elephant in the room, I hope these brief comments explain why.

The Unthinkable and the Impossible

> For an event to take place, for an event to be possible, it must be, as event, as invention, the coming of the impossible.
>
> —JACQUES DERRIDA, *Negotiations*

> Be realistic: demand the impossible!
>
> —CHE GUEVARA

Even if history had a logic the way Hegel or Marx believed, it would not help us anticipate the future. Nor is there a crystal ball giving us special access to what is to come. And yet no thinking person who reflects on what is being said about climate change can avoid constructing hypothetical chains of inference about the next few decades. Claims to know the future are foolish, but the refusal to speculate, to ask "What if . . . ?," is inexcusable.

I propose here various scenarios, consequent on serious climate change, some apocalyptic, some not. My argument will be that to avoid the unthinkable, we will have to imagine the impossible, indeed more than "imagine" it, welcome and embrace it. Moreover, as Heidegger would say, if we cannot bring about what we seek, we can, in some respects, prepare the way. In laying out these scenarios, I will of course be making all manner of assumptions. Some I am aware of, some not. We are undoubtedly in that Rumsfeldian condition in which "There are known knowns. These are things we know that we know. There are known unknowns. That is to say, there are things that we know we don't know. But there are also unknown unknowns. There are things we don't know we don't know."

I begin in familiar territory with the dire warnings we have been given. Of course, I delight in the Beyond the Fringe skit about the End of the World! "Not quite the conflagration we'd been banking on. Never mind lads. Same time tomorrow. We must get a winner one day." I too smile patronizingly at people wearing "The End Is Nigh" sandwich boards. And I am even tempted to display the bumper sticker "When the Rapture Comes Can I Have Your Car?" No doubt the long history of apocalyptic and millenarian predictions reflects social and personal anxieties. Comic relief feeds delightfully off these worries, diffusing them for a time, even as it leaves a troubled residue of concern.

It is true: We cannot just put our faith in "science" (including the "science" of climate change), when no one science can offer a grasp of the whole. We are rightly skeptical about predictions based on simplified models, in which, typically, linear trends are extrapolated indefinitely. For example, it seems there are many positive feedback effects that are accelerating global temperature rise faster than originally predicted. Melting ice lowers reflectivity of Earth's surface and allows more heat to be retained by the darker ground. And in Siberia, vast areas of tundra are moving out of permafrost releasing trapped methane, an even worse greenhouse gas than carbon dioxide. Against this, there may be unexpected countervailing forces, which would dampen climate change.

Nonetheless, I have been convinced (as have many official bodies) by the reality of the threat of climate change. I take seriously the catastrophic consequences of a 2°C rise in average global temperature, and the real prospect of our exceeding that. We are already well on the way—0.8 percent in 2017— and should get there by 2028.

The likelihood of such a rise in global temperature (even in the face of clear warnings) can be explained in many terms: vested corporate interests (funding disinformation through think-tanks), the political difficulty of prioritizing climate change initiatives in the face of the immediate need for jobs and so forth, public unwillingness to make sacrifices or lifestyle changes in the absence of any clear and present danger, the conviction on the part of rich and powerful individuals and nations that they can insulate themselves from the impact, the refusal to believe that God would let this happen (or that Jesus would not return to fix things), hostility to "science," the lack of any sense of agency especially when traditions of political organizing have withered on the vine, and the difficulty of mobilizing in the absence of an external enemy.

I take seriously the message of Bill McKibben's 2012 "Do the Math" lecture tour. If China, India, Brazil, and other developing nations approach the American standard of living, we will require four Earths, in terms of resources and sinks. McKibben also calculates that fossil-fuel companies (especially oil and gas) are sitting on five times the reserves required to exceed the 2-degree mark (which some say is already too low an estimate). Using the reserves would release 2,795 gigatons of CO_2. At under 565 gigatons, we breach the 2-degree line. After that, as James Hansen says, "It's game over for the planet."[1] What should we make of that? Do those CEOs, legislators, decision makers know something we don't? That we will just keep on truckin' even after leaving the edge of the cliff?

The point of this preamble is simply to frame and motivate the argument. It proves nothing, except perhaps that climate disaster is on the cards, even if we have (and can have) no proof of when, how bad, and for whom? Half the nonhumans on the planet wiped out? Africa becomes uninhabitable? Does this matter? And to whom? I am working on the assumption that when the IPCC says going over 2°C would be catastrophic, it is choosing its words carefully. We know it covers threats to the food supply, water supply, air quality, plant life, mass species extinction, and extreme weather. The reversal of the Atlantic conveyor would turn northern Europe into Siberia.[2] These are the direct effects. The indirect ones would be crop failures, famine, disease, war, forced migration, and so on.

A Stark Choice

If one accepts this premise, what is to be done? The solutions, I suggest, are either unthinkable or impossible. Let me explain what I mean.

By unthinkable, I am pointing to one of the following scenarios, each of which would arguably bring relief to the planet as a whole:

1. Allowing whole populations to starve to death, or die of disease, or kill each other.

2. Imposing a new slavery on subjugated populations—enforcing minimal energy consumption, and carefully regulating reproduction rates in line with production needs. This could happen within nations, or between nations, or a mix of both.

3. Military intervention—wiping out whole nations, cities (neutron bombs), or races (gene-based weaponry).[3]

4. A global euthanasia program—either by force, by coercion (money, privileges—as in jail). Or by irresistible inducements—offering a drug induced high for a week, and then . . . Aka the "Die and go to heaven" deal.

5. A global sterilization program—as in (4).

This list of unthinkables is limited only by our imagination. They all assume that reducing the global human energy footprint is good for those remaining, and that this can be achieved by reducing the population, the standard of living, or both. I call them unthinkable because accepting that we live in a world in which they are real options is deeply dispiriting. They are in that sense "unthinkable," spelling the end of progress, the enlightenment project, the belief in a just world. Perhaps more frightening is that "we" (in some sense of "we") have already witnessed local outbreaks of these unthinkables. Biopolitics offers a well-stocked larder after Auschwitz, after Hiroshima. (1) In the 1994 Rwanda genocide 800,000 Tutsis were killed, and women were systematically raped, as they have been in war zones all over the world. (2) North Korea (and parts of China) currently operate a slave worker economy. (3) Both Russia and the United States have developed genetic weapons, and the United States has contemplated the deployment of neutron bombs. (4) "Die and go to heaven" may be unthinkable, but it is graphically conceived in the movie *In Time* (2011) where the poor die young, and the rich can live forever. Money is time. And one could even imagine a voluntary program in which one would be given a week's ecstasy (like a religious experience) before ascending to heaven. (5) Forced and "voluntary" sterilization were once normal in the United States and widely practiced elsewhere for some sex offenders, the feeble-minded, and women who violated the "one-child policy" in the case of China.[4]

The upshot of this charter of horrors is twofold: First, to lift the veil a little, to show that the seeds of the unthinkable have already been sown. We may say of the Holocaust "never again," but in doing so, we take our eye off the ball. It's not as though new forms of horror would be more acceptable. Humans are not to be understood just biologically. Population pressures, economic hardships, desperate times—not to mention greed and lust—have consistently motivated the most awful events: war, persecution, enslavement, incarceration, torture, and genocide.

Rats trapped in a box and without food will kill and eat each other. For all our high-mindedness, history suggests that peace and cooperation are threatened by scarcity. And climate change promises scarcity (and worse) on every front. The alternative to addressing climate change is not carrying on as normal; anything less than serious solutions will mean sliding into a world where the new normal echoes old horrors. It is unfortunate that many in the West do not acknowledge the kind of suffering that could be around the corner. America did not learn from Hurricane Katrina in New Orleans, where in August 2005 some 80–90 percent of the population was evacuated, 1,500 people died, and poor African American communities suffered most. Many white people viewed it as if it were happening in another country.

The Call of the Impossible

Faced with the real prospect of the unthinkable, whether in twenty years or fifty, it is time to turn to considering the remedies offered. This is a most delicate step. Many respond to climate change with resignation. They are convinced by the diagnosis but cannot swallow the prescription. I want to steal Derrida's term, "the im-possible," to suggest that the solutions we need are *impossible*, in the sense that we cannot see how they could come about. It is in this sense that Russell once said that seeing the other side of the moon was impossible. It once was indeed inconceivable.

It might be thought that I am already counseling despair. Not so. But the problem we face is not just a practical one, not just a problem internal to our being in the world. Rather it is an existential, ethico-political one, touching on the deepest questions of dwelling, and will require a response of the same order—a revolution.[5]

Suppose we characterize the problem in this way: An increasing number of people around the world understandably want the goods and services associated with our current Western lifestyle, which currently carry with them an unsustainable aggregate carbon cost.

Three Proposed Solutions

To oversimplify, we can then divide up the kinds of solutions into three categories: (1) Reduce the population or its rate of growth; (2) End or reduce consumerism; or (3) Find new technology to enable currently desired goods and services to be supplied with a much lower carbon footprint. Obviously actual positive outcomes are likely to include a mix of all three. But there are immediate obstacles encountered on each front.

1. The right to reproduce is taken to be a basic human right, one enshrined in the UN charter, and there is understandable resistance to governments interfering with it. While birthrates decline naturally with development, the development process swells the carbon footprint. Even as the Asian birthrate declines, the per capita and aggregate carbon footprint grows. Sending excess populations to colonize other planets is just escapist fantasy—as is being saved by the Rapture, which if it happened would indeed benefit both Christians and those left behind.

2. Consumerism reflects widespread human desires, even if much of it can be laid at the feet of the images of Western lifestyle happiness being broadcast into the homes of the developing world. Renouncing these desires is hard both for those used to them and for those who aspire to them. It strikes at the heart of who we think we are and what we want from life. And to the extent that our economies are predicated on growth, and growth is tied to consumerism, short-term anticonsumerist attitudes would bring economic stagnation and political failure.

3. There is something very hopeful about technological salvation. It appeals to our sense of the unlimited creativity of the human mind. It often bypasses the need to change our basic mindset. And there are undoubtedly success stories out there—"cradle to cradle" waste-free manufacturing design (McDonough)[6], public transportation, microwaves, the Internet, cell phones, solar power—each of which could be said to cut carbon emissions directly or indirectly. But there are also many counter-

vailing considerations. The technology that heralded the paperless office led to the consumption of more paper. The research that gives us electric cars may slow down the move to better public transportation, and be a net negative. Skyping may replace some conferencing, only to introduce us to more people we would like to fly to meet in person. Faith in technological fixes may blind us to the need to make changes of a different order. And finally, if we put our faith in technology to square the circle, we would have to wait a long time. We do not have a long time.

If it is indeed possible that there is no single technological fix, or group of them, we would be mad to put all our eggs in that basket. And frankly, we cannot know ahead of time. Technological fixes are not impossible. In some ways, they are all too possible; they are within our current paradigm. But they are not guaranteed. So, the precautionary principle would require us to look to one or other of the first two categories—population and consumption patterns—for at least part of the solution. Here is where the impossible comes into focus. And unless I have missed something, they will be impossible but necessary.

Reproduction

We can address questions of reproduction and consumption from opposite ends—that of the broader economy (the "growth" economy) and the economy of desire (the shape of what we think we want, who we think we are, and why). In some respect, there is hope here because the human species is remarkably adaptable in both respects. Western-style, consumerist individualism may be the dominant paradigm today, but it is far from being the historical norm. That it seems to set the bounds of foreseeable possibility is what makes imagining change impossible. This is particularly hard to imagine given that the trend in developing countries seems to be away from sustainable family structures and consumption.

The money economy is clearly linked to the economies of identity and desire. Declines in birthrate are linked to the empowerment of women, correlating with contraception and social safety nets that make it less necessary to have children and partners to provide for one's old age. And in a consumer society, rearing many children is expensive—paying for food, clothing, health care, and education. It blocks women's entry into the workforce. But this very individualism also reinforces single-family homes in which bearing and raising children is partitioned into units that forsake the benefits of shared households.

Shared households provide one answer to how couples could have even fewer children without losing the benefits of larger families. At stake, of course, are questions like: Is this *my* child? Do I need a son, a daughter to identify with to perpetuate myself? What would it be like to share children? Who is responsible? And so on. It may seem that such families would not catch on. But adoption already breaks the requirement for genetic identity bonds, and vertically extended families used to be common. Who knows what could be accomplished by tax breaks and other incentives?

Consumerism

Let us, however, focus more explicitly on consumerism. Recall the Brundtland Report *Our Common Future* (1987), which defines sustainability in terms of needs: "Sustainable development is development that meets the needs of the present without compromising the ability of future generations to meet their own needs." Consumerism, however, is not about meeting needs; it is about satisfying desires, preferences as economists would put it, without regard for where these desires or satisfactions come from or lead to. From the classical liberal perspective, any political critique of desire is an unjustified intrusion on individual freedom. Who are you to say what I should want?

But for classically liberal theorists (such as Mill), desire satisfaction *is* an issue if it harms others. And it is down such a path we must go. We must realize that harm may come from the general shape of a practice (rather than a specific act that harms a specific person), and that harm extends to non-persons. This enables the development of legal constraints on individual actions, even if the scope of such constraint is disputable. (For example, taxing personal carbon emissions might meet resistance.)

No doubt there are many ways of working within an individualist consumerist perspective to generate environmentally friendly incentives. But the speed and political acceptability of such incentives will depend on public sentiment. And there is every reason to think that public sentiment is a lagging indicator. What MacPherson called "possessive individualism" treats inroads into its reign as threat and loss. It was in this light that the authors of *Rebuilding America's Defenses* wrote, "the entire Middle East must be reordered according to an American plan." PNAC's most important study notes that selling this plan to the American people will likely take a long time, "absent some catastrophic catalyzing event—like a new Pearl Harbor."[7]

9/11 proved to be such a "catalyzing event," legitimating the invasion of Iraq and everything that followed. Importantly for our thesis, what followed was a national security clampdown (airport security checks, terrorist databases, breaching international law), which reinforced a strange union between a reactive, anxious individualism and the expansion of panoptic state surveillance.

Yet Another Pearl Harbor

Some have suggested that we need an environmental Pearl Harbor to grab people's attention, such as the collapse of the Antarctic Ice Sheet 14,000 years ago, which raised sea levels by forty-five feet. But that would be like waiting for the first tumors before you stopped smoking. And there is every reason to think it would be used to justify draconian emergency measures only nominally connected to the actual event.[8]

Consider the following: Global warming is the aggregate outcome of millions of people voluntarily doing ordinary things (actions sustained and approved by local cultural norms), few of which we could call wrong, let alone evil. Industrial effluents (massive oil spills, toxic waste dumping, nuclear accidents) are very real, but that we could probably cope with them were it not for the impact of mass consumer practices. Of course, they are not really separable. We can drive our cars to the store only if somewhere fossil fuels are being extracted from the ground, transported, refined, and so on. Even then, we are largely not dealing with evil, but with usual business and only occasional serious negligence. The same might be said of warfare. It does not require evil people. But it does require devising efficient ways of killing your fellow man, which cannot in itself be a good thing. My point is that the release of greenhouse gases does not require lethal intent, or psychopathic attitudes, even though it may destroy us. The change that is needed is one in which we alter our behavior without being able to say that any one example of it is wrong. The usual ability to draw on traditional sources of disapproval is missing. To many of our normal practices, tradition would give a thumbs-up. And there is no obvious external enemy.

What these examples demonstrate, I believe, is that biopolitics operates precisely at the level of identity and desire—particularly focused on channeling anxiety into justifying security measures. Desire here involves the acceptance of authority, and identification involves the state (patriotism, support the troops, etc.) and uniting in the face of a common enemy. The reactive form of these forces is alive and well. But how helpful is this when thinking about consumerism?

For the Stoics, consumerism is a pattern of false desire. You only think what you want will make you happy, but when you get it, you remain unsatisfied and immediately redirect desire elsewhere. (Good desires, by contrast, lead to satisfaction and happiness.) And even though false desire is productive for jobs because it creates an insatiable demand for new products, it is unsustainable because it constantly raids the piggy bank of natural capital in resources and sinks. Indeed, numerous studies show that consumerism does not make people happy. At best, you impress your neighbors, gain prestige, and place self-worth on shiny possessions. But this widens the hollowness.[9] Consumerism rests on an illusion that commodities can substitute for what really makes people happy: love, friendship, family, community, real security, good health, and freedom from hardship. It is essential that consumerism never quite delivers on its promise of happiness.

Beyond Consumerism?

The question is whether there are viable alternatives to consumerism. Is it possible to twist free from its "economy," both psychological and material? Or is this precisely the most powerful figure of the impossible?

Here I resort to shorthand. From Marx, Nietzsche, Freud, Heidegger, Irigaray, and countless thinkers of the twentieth century, we have the ingredients for a multileveled diagnosis. Questions of love, desire, identity, and mortality are intimately connected. And they are connected to the kinds of relation we have to our productive labor, our social relations, and indeed our sense of connection to the natural world. Much thinking about patriarchy, for example, argues for its impact right across the board (suggesting recurrent, dysfunctional patterns of relation), including the domination of nature. And perhaps that's the point. There are many shapes, economies, and forms of life that more or less work under limited conditions, but, when pushed to an extreme, break down. Consumerism is one of these. Constant substitutions are psychologically unsatisfying, and they bleed nature dry. The tradition I have gestured toward consistently proposes alternatives to our alienated condition, none of which ever seem to catch on for more than a privileged few. Why should things be different now?

Take the case of Wendell Berry, for example, probably America's preeminent agrarian poet and essayist, author of the classic *The Unsettling of America*. He is an acerbic critic of big agriculture (indeed big anything) and a champion of the small farmer, community, and the local in every form. All this is coupled with a disdain for consumer culture and its false values. His critics say that this authentic agrarianism has little to say to urban America. Few of us are faced with the question of whether to plow with a tractor or a horse. Isn't Wendell Berry just dreaming of a long-lost world to which we will never return? But he is not saying that you have to be a small Kentucky farmer to be happy. He is a practitioner of virtue ethics, championing certain kinds of values, specifically those that flourish in local communities: trust, humility, conversation, patience, and, in the tradition of Aldo Leopold's land ethic, respect for land and place and being able to take the long view. These can be translated in principle into any context. An economic reductionist might doubt that these values could easily take root in a competitive urban economy. Berry's bet, which is not unreasonable, is that we can edge toward the kind of values that would reflect a more benign social existence, especially when we *know* that things cannot go on like this. We can try to be the change we want to see.

There are those who advocate alternative economies. The degrowth movement is one of them. They correctly insist that if continuing economic growth is unsustainable, we need to devise economies that do not depend on it, with the values and virtues that entails.[10]

Shifting Identity Boundaries

What has been dubbed "nature deficit disorder" may best be treated by a good hike in the mountains. But the deeper value of hiking is that brings us closer to the natural world. What that means is seeing oneself as part of the natural order, rather than as some sort of privileged exception. It means identifying natural processes operating within one's own body—even aging and death—and coming to terms with that. One comes to see one's fate, and that of fellow humans, as bound to the fate of nature both morally and materially. This would mark a different mobilization of identity boundaries from that of post-9/11 patriotism, but it would be playing with the same ball. What am I part of? What do I truly belong to? What and who, if anything, belongs to me, to us? It would be easy to understand such a shift in terms of renunciation (of mastery, for example). But there is also a certain pleasure, even joy, in a new intimacy of connectedness.

Such a shift in identity-boundaries could even recalibrate our being-toward-death, the importance of which can hardly be overstated. As Robert Pogue Harrison put it: "when we do not speak our death to the world we speak death to the world."[11] The impossible, the Derridaean im-possible, arises here because such new attitudes do not make sense in the old logic. Letting go is not a move in the game of control. Or is it?

Our desire for control is out of control. The "I" that seeks control ceases to be sure of its own ground, perhaps glimpsing its dependency on what it cannot control. There is an analogous point in the later Heidegger, when he writes about ceasing all overcoming and letting being be, twisting free of the previous logic. Similarly, in *Contributions to Philosophy* Heidegger sets aside the temptation of new, willful, programmatic action in favor of "preparing the way," so as not simply to reinforce the logic of agency at the moment of its fundamental critique. Again, we are in the territory of the impossible.

The trouble with these kinds of moves—opening oneself to impossible possibilities—is that such an attitude does not seem to address the actual historical material urgency of our condition. It's a bit too much like waiting for the second coming. Derrida used to take stick for talking about the need to go through the undecidable if we want to make a responsible decision. It sounded like a recipe for interminable delay. No, he insisted, it is true even if we should have acted yesterday; not actually having time to act "responsibly" does not change what that would mean.[12] Can a similar response be made here? I'm claiming that we have to choose between an unthinkably awful future that business as usual portends and a dramatic realignment of our ways of life. Not only is there deep resistance to such a realignment, but also to the reality of the stark choice. The British expression for what we would like to think we can do instead is "muddling through." This might be thought to be utterly irrational. But it could itself reflect a generally healthy disposition to mistrust such dramatically constructed options. And, indeed, it is often the case that circumstances change, fixed points turn mutable, and new possibilities emerge unexpectedly. A meteor might strike the earth, decimating life as we know it, just in time to head off the agonizing choice.

On Not Just Muddling Through

But equally it might be that while the "muddling through" attitude works as a rule, it makes no sense in this situation. An airline pilot, using such words to dismiss warning signs in his cockpit, would not inspire confidence. Nor would a brain surgeon, breezily waving his scalpel in mid operation. And we cannot think we can split the difference between those who think that $2 + 2 = 4$ and the $2 + 2 = 5$ers. These considerations take us back to our insistence on giving Nietzsche's critical take on history some privilege. While it may be rational to generally mistrust simplistic choices, we also need to suspend such skepticism when appropriate. If in fact we do choose the impossible, it would be the rational thing to do, not least because we are and should be terrified by the unthinkable alternative. Such terror would be a rational passion.

However, if we took seriously the mood of Heidegger's *Contributions*, we would have to step back from a simple model of "choosing." "Preparing the way" instead reflects both the difficulties of constituting an appropriate collectivity for action (which "we"?), as well as the extraordinary barriers to changing the background conditions (such as consumerism, free market capitalism and so on) that are for most of us most of the time not items of choice but the very space of choice. If responding seriously to deep time requires a geological consciousness, part of what that entails is grappling constructively with the limits of our power to do anything. The hope must be not that a messiah will save us with a grand gesture, but that new possibilities of agency, and new ways of constructing identity and desire will usher in the change we need. These cautionary observations must be borne in mind when in Chapter 10 we ask again: What is to be done?

What Is to Be Done? Democracy and Beyond

Democracy is the worst form of government, except for all the others.

—WINSTON CHURCHILL

The Republican Party is the most dangerous organization in human history.

—NOAM CHOMSKY

The overarching idea of this book is that we humans find ourselves with a geological-scale awareness of the history of the planet, indeed the cosmos, and a capacity to project its immediate and long-term future. At the same time, we realize that our own activity is destabilizing the ground conditions for the evolutionary processes that make us possible. The "we" that is having this impact is a mass phenomenon, not currently subject to collective direction or control. In the absence of such control, a change of course seems unlikely, and disaster highly plausible. How are we to think about this situation?

Homo sapiens has been around for perhaps a couple of hundred thousand years. While what we call democracy may well build on collective decision-making strategies already found in other creatures (electing a "leader"), it may be our single greatest achievement in this arena, and yet it seems to be failing us. So it could be argued that we need to expand and perfect democracy.

Difficult Democracy

But democracy is not without its difficulties. Consider some real-world limitations and cautions. First, there are plenty of nation states that do not enjoy democratic institutions. It is unclear how much impact the advice to expand and perfect democracy would have here. The promise of delivering democracy to less politically developed parts of the world is a notorious alibi for economic colonization by states whose own democratic credentials are increasingly suspect. Second, given that nonhumans are stakeholders in the earth enterprise, we need to find some way of representing their interests, or giving them a voice. The same goes for actual and possible future generations, both of humans and nonhumans. Third, international agreements, needed for tackling global issues, are typically subject to veto (for example by the UN Security Council). There is little international democracy, and no "agency" effectively representing the health of the planet, let alone taking a deep time perspective.

Even setting these limits aside, fundamental objections to democracy date back at least to Plato. First, there is an in-principle objection: there never was, or can be, "real democracy." Democracy is a vague aspiration, inherently flawed as a premise. Second, there is a transition problem: How can we get from our real world to a more ideal one, from our current deeply compromise system to true democracy? Third, populism should worry us. Democracy can and does legitimate the tyranny of the majority over the minority. There is nothing to say that the victors in a democratic process have to consider at all the interests or views of the 49 percent who did not vote for them.

But one objection deserves specific attention. Human beings are irrational and easily manipulated. Concerning rationality, there is no necessary connection between a policy that captures public passion and the public choosing their best interests (let alone one that addresses the interests of noncitizens or nonhumans). Representative democracy "works" (at least formally) when it represents people's views, however misguided they may be. As for manipulability, there's no point in making the machinery of democracy run smoothly, without simultaneously changing hearts and minds, and cultivating an educated citizenry. Those who fund disinformation think-tanks to corrupt the public sphere know this all too well. It is not idle speculation to suppose that a democracy could vote to act in a way that would most likely bring about a catastrophic outcome (such as severe, irreversible climate change). That is what lies behind Chomsky's "outrageous but true" claim about the Republican Party. He believes they are licensing the destruction of the planet.

People repeatedly vote against their own interests, often because politicians or newspapers lie to them or make empty promises. Representative democracy, driven by elections, only delivers just goods with informed citizens. Moreover it is not enough that they be informed. For justice to be served, they need to care about those without a voice—other nations, and other creatures. In the case of no-return climate change, this matters. Things can't be put right next time around if there is no next time, or if it is too late.

A Thought Experiment

Suppose then you believe in democracy. You believe, with good evidence, that it would take at least four years to potentially reverse a vote that exacerbated climate change, and that such a delay would be fatal. You never believed that democracy would necessarily yield the right decision, but all mistakes before this one could be corrected (in some sense) on the next cycle. You still believe there are good general procedural reasons for supporting the democratic process, reasons connected to the fragile achievements of social justice, healthcare, education, civil rights and so on. As it happens you have the power to subvert the will of the people and head off the catastrophe. Perhaps you secretly have Gyges's invisibility ring and can use it with no negative consequences. Would you subvert democracy?

I am assuming here a "reasonably fair electoral procedure"—free from some or all corporate money, voter suppression, foreign interference, blatant lies, gerrymandered redistricting, and so on. (If those assumptions *cannot* be made, the question gets even more pressing. The legitimacy of a sham democracy would come to rest not on the principles it embodies but on the sorry idea that even the illusion of democracy can provide a ritual basis for heading off internecine conflict.)

This thought experiment contains a practical, strategic question about ethics and politics, but also an existential one. Would you use your magic ring to bypass democracy if it was the only plausible way to head off the End of the World? Many people would say yes. This is obviously a comically simplistic thought-experiment. But history has offered scenarios that are not so different. Dietrich Bonhoeffer, a pacifist theologian, lent his support to one of many attempts to assassinate Hitler. The stakes were that high.

The positive response to this question is in line with Schmitt's approval of "an exceptional rejection of democratic principles." His "state of emergency" argument is meant to justify temporary dictatorial power if needed to preserve the state, or even the prospects of a future democracy.

Revolution

A plausible argument can be made that links the failure of "the democratic process" to seemingly unstoppable climate change. Democracy works at best with shallow rather than deep time (think electoral cycles). And if such change really would be disastrous for anything like democracy maintaining a foothold in our political life, then, in the absence of a magic ring, mass resistance, even violent revolution could be justified. This is a step beyond Earth First's strategy of monkeywrenching—the destruction of property and machines used in felling old growth forests and the like while being careful not to harm humans.[1]

Frustration with democracy is endemic even among those who formally enjoy its institutions and rituals. It's not that they would prefer a dictatorship, but that they are deeply disappointed by how far and how often democracy falls short of its promise both in process and outcome. A better world seems so obviously possible. But even if revolutionary action could in these circumstances be justified, one would have first to believe that it had some chance of success. The briefest account of why I do not think that any classical understanding of revolution is viable today, at least in the West, is that it involves "seizing power." This requires believing that power can be identified and targeted, in ways that are no longer convincing. One could take control of the machinery of state but not financial markets, not the media, and not the multinationals. And even if power could be seized it would be worthless without a change of heart among the general populace. If this argument is sound, the implication is not that democracy needs to be overthrown, but that we need to step sideways to address its deficiencies.

Economic Solutions

Representative democracy is not the only game in town. Many who have lost faith in the political process believe that consumers and corporations can bring change by the choices they make independent of government public policy. Corporations can make profits by going "ecological." People can voluntarily buy green products and change their behavior. And here, in a strange way, some of the very forces that sabotaged voting—that people will vote their identity rather than interests—can play a positive role. To affirm a green identity, people will buy goods that reflect or confirm that identity, against at least a narrow view of their economic interests. They will choose environmentally friendly products, even if they cost more.[2] Such behavior provides another way through the collective action problems that otherwise plague instrumentalist understandings of agency. My doing the right thing may not help save the world unless many others do likewise, but it can make me feel good. Such "feeling good" operates at the level of identity, self-understanding, and self-worth—powerful motivators for good or ill. The upshot of this argument is that, whether or not one believes in the efficacy of formal electoral democratic processes, a certain green awakening is a worthwhile goal. It might also help elect different people or empower elected officials to act in ways they otherwise might not, and it will help direct consumption in eco-friendly directions.

Passions, moods, and basic dispositions matter, and each of us packs a full palette. But it is arguable that economic conditions occasion their selective expression. Couple this with the way in which job creation is routinely used as an argument to suspend or cut environmental protection, and another imperative emerges. If our economic system needs fewer people to produce the goods, but still needs them to buy the goods, and if people need jobs both to be healthy consumers and to be less susceptible to regressive political propaganda, then we need to promote new forms of employment. In particular we could especially promote personal services that maintain a developing social fabric with minimal environmental footprint. This is not an argument for more servants in a trickle-down class hierarchy. The new shapes of such services can arise spontaneously—personal trainers, counselors, dog-walkers, and so on. And we need socially oriented think-tanks to reimagine work, and even social relations. For example, it has often been remarked that not having a child or having fewer children is for most of us the single most significant contribution we can make to cutting our carbon footprint, dwarfing all our other eco-friendly gestures. For many others, not flying to conferences or to visit relatives is the next most effective contribution. Followed by giving up meat. However, especially in the affluent West, each of these suggestions can be seen as an attack on maternal instinct or the basic shape of our connection with family and friends, or with community and tradition (barbecuing, and so forth). Such changes are vehemently resisted, even by otherwise progressive friends. We may lament the reckless crippling of the Environmental Protection Agency and the loosening of environmental controls, spurred on by those angry at being held back by "Washington" or tangled up in red tape. But attempts to limit family size, curtail recreational travel, or tax meat would be met with anger, defiance, and outrage.

There is no point, one is told, trying to sell a negative message about our environmental responsibility. (Compare Nietzsche's complaint about the burden of history.) If however you are in fact faced with a choice between the unthinkable and the impossible (see Chapter 9), the proper question would seem to be: Would you rather risk climate catastrophe or fairly serious changes in your current lifestyle? We are too used to the idea we can have our cake and eat it.

With these considerations, our focus on the affective and its broader frame (moods, dispositions, modes of attunement) regain their focus, and their connection with deep time should perhaps be restated. Our palette of affective responses is the product of evolutionary history, in which individual and group reproductive success and survival have been the selective mechanisms. But the circumstances in which our passions organize themselves to promote survival and flourishing have changed. Environmental pressures are not local but global. And the social and political institutions, by which our interests and passions are collectively steered, have not kept pace with the challenges that face us. Time is out of joint. Normal mechanisms of self-correction by which mistakes are rectified the next time around are not adapted to the prospect of irreversible changes in climate. The temporal dimension of the crisis erupts too in the struggle between the effective drag of everyday time: the way habit, custom, ritual, business-as-usual ease the slow slide toward the precipice of an uncontrollable climate. There is what could be called an affective discount rate. We are not wired adequately to weigh unprecedented but predictable future disaster, when faced with the alternative of disruption to today's sedimented habits. "We'll cross that bridge when we come to it." Such flawed metrics may have made perfect sense in times when personal survival even from day to day was not to be taken for granted.

Our consideration of the economic dimension of our current crisis has been largely limited to the acknowledgment that economic conditions clearly can drive the public mood, which in turn can set the scene for regressive political choices that close down opportunities for the changes needed. Marginally we have noted that to the extent that green consumerism feeds on identity concerns not reducible to strict economic rationality, it offers a way in which we might begin to buy our way out of trouble. But there is another source of economic breakthrough worth noting. Paul Mason argues that free information disrupts economies based solely on market mechanisms. The information freely available on the internet (through Wikipedia, Creative Commons, and Open Source) makes possible to take college level courses, publish your own book, create your own home business, connect with global shared-interest communities, and share images and information.[3] Is this an economic revolution? It has been convincingly argued that market economies are not, as such, the scourge of the planet; what matters is how those markets are allowed to operate. Carbon trading is a market mechanism after all, though it may be flawed, and markets can be regulated. Nonetheless, the idea that freely flowing information inaugurates a post-market economy might well open up new ways forward.

Soft Technology to the Rescue?

There are many proposed hard technology solutions to inexorable climate change out there, notably geoengineering. I have given reasons not to rely on them, even if we might one day strike lucky.[4] I want to end this chapter by opening a different speculative space, one that might bridge the gap between urgency and openness. I also take further Mason's ideas about the promise of the Internet. If identity is linked to relationality, it would hardly be surprising to learn that the internet was affecting identity. Facebook, for example, opens possibilities, unparalleled in human history, for personal connectedness with spatially distant others. It has even spawned a new sense of "friend" and a new verb, "to unfriend." One can carefully construct the identity one wants to convey by the selective release of information. At the same time, there are opportunities for impersonally conveying personal information in which it seems public/private boundaries have radically shifted. Does social media reinforce or deconstruct individualism? We can construct rich virtual communities, but how "real" are they? I do not for a moment discount the well-documented negativities here.

Clearly Google is an information revolution. Knowledge of what is going on behind closed doors, or on the other side of the world, is of enormous value to a global environmentalism (such as the repeated failure of large dam projects, and the corruption associated with them). Does not the Internet presumptively and eventually to serve the truth? Surely the truth will out as never before. If global warming could only happen under the radar so to speak, does not the information revolution arguably prevent that from happening?

Three Pathways

I have suggested three ways in which the information revolution opens up pathways that allow us to bypass the limits of our formal democracies: the formation of virtual communities, the global sharing of information of wrongdoing (and ways of organizing to expose and protest it), and overcoming collective action difficulties. Each of these appears first as a positive expansion or extension of traditional pre-Google powers and capacities. But it has become clear that they each operate in a more complex space that compromises their initial sparkle. Opportunities for community formation among the previously isolated make it easier for hate groups to come together. Corporations and security agencies intent on resisting protests against their destructive activities can and do share techniques of infiltration, harassment, intimidation, and so on. Consider too how right-wing think-tanks and governmental and military agencies now sponsor propaganda groups to control messages, selectively leak information, and attack science? Or broadcast memes tested to reinforce what we already believe? Think of how identities (interests, allegiances, preferences) are being tracked through cookies on our laptops, for both commercial and political purposes. Or how websites can be closed by denial of service attacks. Opportunities for monitoring, personal harassment and corporate blackmail have proliferated. Given the rise of "fake news," are we not being naive in declaring that truth will out? Might we not just as easily find ourselves in a post-truth world, one in which there were *only* truth-claims but no actual truth. And no definitive way of navigating through the jungle of lies, distortions, self-interested exaggerations, earnest declarations, hard science, and sincere witnessings. There are two responses to this. The first is that nothing has changed. It's just that what has been the case all along is now clearer: We do not have access to Truth, but only competing claims to truth. Recognizing this is a good thing. The second response is that while this requires more than ever a critical skepticism over claims to truth, a certain cynicism is now the norm, especially among the young. While there may be no such thing as pure information, the sheer quantity and detail of information at our fingertips is unprecedented. It means we can compare sources, assess contradictions, and take some reports with a pinch of salt. Clearly the debate continues. If it is too simplistic to say that truth will out, the hope is that the googleable world is cultivating the critical capacities that make that eventually more likely. I am not arguing for naïve optimism here, but that the information revolution is one of the most potent sites of discontinuous transformation. Only that will suffice, even if the danger is no less clear.

Collective Participation

But there is another powerful facilitation that the Internet (here social media) makes possible. There are many things that I would do if I thought other people were doing them too, or would do them if I did, but I do not do them because my doing them alone would make no significant difference. The Internet allows precisely such collective reassurance. One can, for example, donate money to an enterprise that will be refunded automatically if the enterprise does not get enough takers. This *solves* the problem of the isolated actor. It also enables the formation of flash mobs, letter-writing campaigns, petitions, and other forms of collective organization. Given that one of the major impediments to changing behavior is the uselessness of doing this one at a time, the emergence of electronically facilitated informed collective conditional participation would be as significant to social change as was the development of contract law to the development of commerce, first in Rome and then in the rest of Europe.

When I first contemplated these matters, I had come to see the Internet as the technological realization of what Teilhard de Chardin called the noosphere, which would be to the biosphere what the biosphere was to the inanimate material world. It would mark the birth of a new spiritual dimension to terrestrial existence, made possible by the buzz of human communication and collective engagement. I am cautious about the sense of a radical separation from the natural order that this implies, but the idea of a new band of communicative intelligence, coupled perhaps with a growing sense of terrestrial responsibility and cybercitizenship, is not impossible to imagine. At the very least it allows us to toy at least with the development of a new sense of identity, imagining the impossible. Whether it could contribute to the withering away of consumerist desire is another question. Wendell Berry would doubtless shake his head at the very suggestion.

The logic of the argument here is this: with respect to climate change, carrying on with business as usual will likely lead to unthinkably awful outcomes—holocausts of negligence. These outcomes would not just be dreadful, but they could destroy all hope. If this is to be avoided, there are only a limited number of options.

While technology might come to the rescue, the belief that it surely will is no better than faith-based religion. The various remaining options center on transition to a post-consumerist culture and economic order, for which we have only the sketchiest of blueprints. I have argued that its seeming impossibility rests on it being tied up with questions of desire and identity, the current forms of which are already unsatisfying and ripe for transformation, and that the contemporary explosion of communication media opens a path for this to happen.

We find ourselves with some strange futurological bedfellows. First, we mentioned Teilhard de Chardin's noosphere breaking free of its biological base. The autonomy of intelligence is in fact another myth, potentially dangerous, but the idea that the natural causality might be importantly supplanted or even supplanted by information flow cannot be discounted. And the hope of moderating the atavism that drives much of the darker side of human history—especially war—must be taken seriously. So must the thought that it offers a path to a more altruistic model of identity (fragile, and not unproblematic—see Deep Ecology and ecofascism). Consider, too, Kurzweil's idea of the singularity, the point at which artificial intelligence will take over from human intelligence. Could this be a metaphor for the possibility of transformations in identity and desire that we have been talking about? We do not need Google hardwired into our brains to recognize the extraordinary leap in information access and social communication that it makes possible. Even as we lament the fact that guests consult their smart phones over dinner, we cannot deny something is shifting. To adapt Bergson, we might call this the age of Creative Connectivity. We may have no more reliable ways forward. It might be our last best hope.

A Measure of Hope

There is doubtless an apocalyptic tone, or undertone, to this book. I have sought however to dispel the suggestion that this tone merely reflects the contemporary mood. I want both to argue that in the absence of obviously available forms of collective agency, a whole range of public responses— resignation, anger, apathy, *ressentiment*—are entirely understandable. (They are plausible responses both to local political failure and to the seemingly intractable prospect of serious climate change.) But I am also claiming, for my own part, that the dramatic narrative to which I am subscribing— impending climate disaster—is (sadly) well grounded, even as explanations for its denial abound.[5] I have not provided chapter and verse for the science behind some of the frightening predictions that have been made.[6] Nor have

I given detailed descriptions of what is in store for us. These are not lazy omissions. I assume that anyone paying attention has read them. Moreover, we do not need to know the details of impending mass-famine migrations from a desertified Africa to want to pull out all the stops to prevent it. A reader who does not accept these assumptions is unlikely to have reached this point in the book. I must also confess a receptivity to the possibility of dramatic change, the unexpected, the world being turned upside down. Again, one could try to explain away such an attitude. But those who have suffered war, tyranny, invasion, and the breakdown of law and order both in recent decades and in the broader course of history, would find it unsurprising. Those of my generation who watched the tragedies of Vietnam, Syria, Afghanistan, Iraq, and too many more unfold before our eyes have only had to imagine at a distance what dramatic change to everyday life is like. My sense that things can change rapidly for the worse, that such changes may be more or less irretrievable, and *that they did not have to happen* is no idle supposition. It is accompanied by a strong appreciation of the power of habit and the shrill demands of the short term. Sometimes even warnings about catastrophe get kicked into the long grass, for good reason. Receptivity to the possibility of dramatic change is further strengthened by reflecting on how this happens at an everyday level all the time: Wine overflows the glass, rain relieves a drought, a dictator type gets elected, a relative dies, a store goes out of business. On a small scale everything changes; tipping points are real. And while all this is happening "in the real world," perhaps not coincidentally I subscribe both to the view that if we humans do not change the manner of our dwelling, we will be forced to face far more unpleasant circumstances, and to the view that the ways we do philosophy can block or facilitate that transformation. The figures I have drawn on—especially Nietzsche, Heidegger, and Derrida—have shared that view, and grappled with its implications.

A Note about Climate Change

The literature on climate change is outpacing even the growth in greenhouse gases. Small forests have gone into the making of these books! They can be sorted into seven categories. I give examples of each.

1. Compilations of the scientific evidence for anthropogenic climate change.[7]
2. Documentation of campaigns of disinformation and climate skepticism.[8]
3. Further explanations of public resistance and inaction (political, cultural, and psychological).[9]
4. Proposals for changes in our practices, and policies.[10]
5. Activist attempts to get us to see how disastrous things could be.[11]
6. Philosophical reflections on the ethical and social justice dimensions of climate change.[12]
7. Analyses of the political and economic root causes of climate change.[13]
8. Religious treatments of the implications of climate change.[14]

While the present book addresses many of these issues along a distinct line of inquiry, a systematic engagement with this vast literature has not been attempted.

A Final Observation

There will be those who complain that my position is deeply anthropocentric. Surely if we are going to take on the geological consciousness of Deep Time, we should be brave and accept Nietzsche's caustic reminder that human life is a mere blip in cosmic history. Should we not be imagining evolution resetting itself after our demise, opening up other perhaps unimaginably glorious futures millions of years down the line? Is my resistance to such a perspective just sentimental speciesism? My response is threefold. First, I believe that *Homo sapiens* is an unfinished project, and that utopian projections of human flourishing are more viable than ever. Second, it is quite possible that there may well be no more interesting or complex form of life in the cosmos than human beings. We should not assume that we believe this just because "we" happen to be human. Third, all the values we bring to bear on the course of history, even geological history, are the prerogative of living beings. Without life there is no value at all. And imagining alternative posthuman scintillating futures may well be the exclusive privilege of the human. Pretending otherwise, I believe, is a kind of alien-ated performative contradiction. I am not denying that we may only have scratched the surface in understanding the distinctive capacities of nonhu-man life forms. Nor am I denying that responsible choices about Earth's future have to allow for the possibility that it might be better off without us.[15] But this latter supposition rests on the promotion of values (like the importance of biodiversity to the flux of life), through which we are affirm-ing our highest concerns. It would not be a contradiction to go down on our sword.

It is tempting to repeat Kafka when he wrote to Max Brod, "There is an infinite amount of hope in the universe . . . but not for us." I am not sure whether hope is a passion, even if hopelessness certainly is. Nor can we be sure we can arouse hope by explaining how it is justified. We would begin by affirming that the world we find ourselves in is a processual one. And while accelerating climate change is the gorilla in that room, there are many other processes—changes in the law, public sentiment, attitudes to social justice, medical advances, communication, social relations, forms of com-munity, technological advances, shapes of employment—each of which are susceptible to nonlinear transformation—tipping points in attitudes, prac-tices and institutions. These changes are not always progressive, but they can be. And they are often unpredictable—think of gay marriage in the United States. The unprecedented possibilities of the Internet and social me-dia really do open a new front. If soft technology puts a new spin on the question as to whether technology can save us, we should not pretend that it escapes materiality completely. It is said that running Google uses as much electricity as the entire nation of Turkey does.

Is this all not a bit speculative? I am reminded here of Derrida's talk of the promise, of messianicity and the to-come. I am allergic to literal mes-sianism, and the invocation of extraterrestrial forces of any sort. I am com-mitted, as was Nietzsche, to "being true to the earth." Derrida points to an openness predicated on there being limits to calculation and predictability. What he called openness to the impossible may be a condition of it happen-ing. And Deleuze, one of the most trenchant critics of regressive desire (and no optimist), imagined a "new earth, new peoples." The logic of my overall argument is this: If these shifts away from consumerist identity and desire are necessary to avoid tragedy, then we need to explore the most promising ways forward. Technology-driven creative connectivity, with all its conspicuous dangers, holds unexplored promise, as well as the more obvious dangers.

If hope demands a clear path to its objective, then it is easy to lose hope. But once we accept that the unexpected does happen, and might well not happen if you and your friends had not acted in progressive ways, then you have a reason to get up in the morning and knock on doors, or join a recycling coop, or turn up at a fossil-fuel company's shareholders meeting, or write a pamphlet. What you do just might all the difference. And most of these less dramatic actions are intrinsically valuable, worth doing anyway.

ACKNOWLEDGMENTS

This book had its beginnings in a conversation with the late Helen Tartar at a SPEP conference. She introduced me to Dimitris Vardoulakis, who invited me to present the 2015 Thinking Out Loud lectures in Sydney, sponsored by the University of Western Sydney. I would like to thank all those involved in organizing this event, and for Paul Albert's formal response. I am especially grateful to interviewer Joe Gelonesi from ABC Radio National, who arranged for the broadcasting of the original lectures, and to Peter Hutchings, Dean of Humanities at WSU, for his support of the series. At the invitation of Alejandro Vallega and Daniela Vallega-Neu, I subsequently reworked this material for a course at the Collegium Phaenomenologicum, Italy, in 2016. Discussion with the participants was especially valuable. Thinking through these issues has benefited from conversations with many friends over the years, including Keith Ansell-Pearson, Andrew Benjamin, Michael Bess, Beth Conklin, Jonathan Gilligan, Richard Kearney, Catherine Keller, Irene Klaver, David Morris, Barbara Muraca, Dennis Schmitt, Brian Schroeder, Charles Scott, and Ted Toadvine. I much appreciate the patience and care with which Dimitris Vardoulakis and Richard Morrison, John Garza, and Eric Newman at Fordham guided this book from the shadows into the sunlight. Amanda Boetzkes and an anonymous reviewer made invaluable suggestions for improvement, for which I thank them. Boomer Trujillo read and helped improve the whole manuscript; Lisa Madura and Eric MacPhail did a great job preparing the index. I am most grateful.

1. HERDING THE CATS OF DEEP TIME

1. Friedrich Nietzsche, "Uses and Disadvantages of History for Life," in *Untimely Meditations* (Cambridge: Cambridge University Press, 1983).

2. There is hesitation in some quarters about the word "Anthropocene." Some argue that it is contested science masquerading as fact. Many dispute the date of its emergence. Some think it suggests that humans are now in charge. Others have conceptual and political reservations about attributing to a species what is the responsibility of capitalism, Western industrial society, or something similar. Yet others understand the word as marking a certain sensibility. Philosophers, it may be said, should be especially cautious in picking up fashionable expressions and using them uncritically. I understand the Anthropocene to mark a substantial and impactful shift in the background conditions under which life has evolved on the planet—such as temperature, weather patterns, ocean acidity, and greenhouse gas levels in the atmosphere—caused by an accelerating level of human industrial activity. It may not be a purely "scientific" concept in that it involves judgments about significant levels of "impact." But the Sixth Great Extinction (of nonhuman species) puts it on a par with other geological-scale changes. The political objection seems to me misdirected. Sure, advanced industrial capitalist production is responsible, but that itself is something our species has developed. And the understandable consumerist aspirations of underdeveloped parts of the world make it increasingly clear that it is now an issue for us to address as a species.

3. Michel Foucault, "Nietzsche, Genealogy, History," in *Language, Counter-Memory, Practice* (Ithaca, NY: Cornell University Press, 1980).

4. Figures include Samuel Alexander, Pierre Teilhard de Chardin, Alfred North Whitehead, and Michael Polanyi.

5. See my "Homo Sapiens" entry in *The Edinburgh Companion to Animal Studies*, ed. Lynn Turner, Ron Broglio, and Undine Sellbach (Edinburgh: Edinburgh University Press, 2018).

6. Historicity (*Geschichtlichkeit*), historicality, and historiography (*Historizität*).

7. See Isaac Asimov's *Book of Facts* (New York: Gramercy, 1991).

8. It is hard to resist subscribing to some version of the critical consensus that predicts climate catastrophe, while selecting judiciously from a list of usual suspects to explain what is going on, even as one knows full well that these totalizing visions reflect interests of various sorts. These include the metaphysical desire for simplicity and presuppose a measured willingness to accept the findings of science (climate science), despite general reservations about the natural-science paradigm.

9. Ted Toadvine, "Deep Past, Deep Future: Anachronicity in the Anthropocene." Unpublished research proposal (2014).

10. David Morris, "Being as Creativity: 'Existent Generality' and Deep Temporality," paper presented at SPEP's annual conference (New Orleans, 2014) and reworked in his *Merleau-Ponty's Developmental Ontology* (Evanston: Northwestern University Press, 2018).

11. Glen Mazis, "The Depths of Time in the World's Memory of Self," in *Time, Memory, Institution: Merleau-Ponty's New Ontology of Self*, ed. David Morris (Athens: Ohio University Press, 2015).

12. See David Krell's work on ecstatic interruptive temporality, especially *Ecstasy, Catastrophe: Heidegger from Being and Time to the Black Notebooks* (Albany: SUNY Press, 2015); see also my previous forays *The Deconstruction of Time* (Evanston, IL: Northwestern University Press, 2001), and *Time after Time* (Bloomington: Indiana University Press, 2007).

13. See John Sallis, *Force of Imagination: The Sense of the Elemental* (Bloomington: Indiana University Press, 2000).

14. Joanna Hodge, *Derrida on Time* (London: Routledge, 2007).

15. Henry Gee, *In Search of Deep Time* (Ithaca, NY: Cornell University Press, 2000).

16. Often left in the original German. It means something like "machination" or "calculated domination."

17. See Miguel de Beistegui, *The New Heidegger* (London: Bloomsbury, 2005), 117.

18. See Martin Heidegger, *Contributions to Philosophy (Of the Event)*, trans. Richard Rojcewicz and Daniela Vallega-Neu (Bloomington: Indiana University Press, 2012).

19. Important companions on the path are Kelly Oliver, *Earth and World: After the Apollo Missions* (New York: Columbia University Press, 2015), and Andrew Mitchell, *The Fourfold: Reading the Late Heidegger* (Evanston, IL: Northwestern University Press, 2015).

20. Edmund Burke, *Reflections on the Revolution in France* (Oxford: Oxford University Press, 2009), para. 242.

21. I discuss his very different account of the eternal return elsewhere. See *The Deconstruction of Time* (Evanston, IL: Northwestern University Press, 2001), part I, "Nietzsche's Transvaluation of Time."

22. See my "Poetics of Time," in *Time after Time.*

23. The challenge of Heidegger's (1962) essay on a primordial time-space is to develop a way of talking that productively reflects this interpenetration.

24. I have visited the torture chamber and felt both relief that the torture had stopped and a terrible ache that it had once happened. Right there.

25. See Michael Shellenberger and Ted Nordhaus, *Break Through: From the Death of Environmentalism to the Politics of Possibility* (Boston: Houghton Mifflin, 2007).

26. The Pleistocene geological epoch lasted from about 2.5 million to 12,000 years ago, spanning repeated terrestrial glaciations.

27. Jonathan Edwards (1703–58) was an American theologian, president of Princeton, and author of *The End For Which God Created the World* and other books. His work was an important part of the Great Awakening.

28. That the world has and does not have a beginning in time (and is spatially finite and infinite): Kant, *Critique of Pure Reason.*

29. The term "negative entropy" was introduced by Erwin Schrodinger in his book *What Is Life?* (1944) and later shortened to negentropy by Leon Brillouin. If "the entropy of the universe tends to a maximum," as Clausius (1865) said, the existence of life has seemed to some a puzzle.

30. Reading Gregory Benford's *Deep Time: How Humanity Communicates Across Millennia* (New York: Harper, 2000) opens up the future in a similar way.

31. Tamsin Lorraine, *Deleuze and Guattari's Immanent Ethics* (Albany: SUNY Press, 2012), 5.

32. Michael Foucault, *The Order of Things* (London: Tavistock, 1970), 387.

33. The subject matter—global climate change—is already multiply temporally stratified. It is the result of releasing exceedingly fast the stored energy locked up millions of years ago from the growth of plants and the conversion of their remains into oil, coal, gas, and so forth over eons. The effects of this release have not all yet shown themselves, and much is still in the pipeline. The reason this matters is that living beings on the planet have evolved, again over hundreds of thousands of years to cope with a narrowish range of climatic conditions, and will not adapt fast enough. Moreover human industrial activity has in many other ways spun out of control, destroying habitats. Developing nations are challenging developed nations to acknowledge their special responsibility and the source of their current wealth and

privilege in their industrial revolutions. And we have not developed the cultural constraints, the social institutions, or the political mechanisms to constraint our own surging development. All of this is set against uncertain futures, timespans, predictions, interests, and so forth.

The Bonn Climate Change conference (2017) is one in a series of urgent conferences on the same topic—from the 1992 Rio Earth Summit to the Kyoto Protocol (2005), the Bali Action Plan (2007), the Copenhagen 2°C limit (2009), with further developments in Cancun (2010) to the 2015 Paris Climate Change conference. The Paris conference was already billed as the Last Chance saloon. Will we finally get it, at a minute to midnight? Or are we inured to tragic repetition as we speed like a cartoon character over the edge of the cliff?

34. The words "man" and "Man" are obviously gender anachronisms today. At times I leave them unmarked in deference to their previous common use.

35. Martin Heidegger, *What Is Called Thinking?* (New York: Harper & Row, 1976).

36. Jacques Derrida, "'Eating Well,' or the Calculation of the Subject," in *Points . . . Interviews 1974–1994* (Stanford: Stanford University Press, 1995).

37. "The sun dates the time interpreted in taking care. From this dating arises the 'most natural' measure of time, the day." Martin Heidegger, *Being and Time* (Albany: SUNY Press, 2010), 303, H412–3.

38. Heidegger, *What Is Called Thinking?* 76.

39. See Jacques Derrida, *The Animal That Therefore I Am* (New York: Fordham University Press, 2008), 3.

40. In Martin Heidegger, *Poetry, Language, Thought* (New York: Harper & Row, 1971).

41. Friedrich Nietzsche, *The Gay Science* (New York: Vintage, 1974), section 341.

2. WHO DO WE THINK WE ARE?

1. See Teilhard de Chardin, *The Phenomenon of Man* (New York: Harper, 2008), where he also discusses the idea of cosmogenesis.

2. Such an account is foreshadowed in Naomi Oreskes and Erik M. Conway, *The Collapse of Western Civilization: A View from the Future* (New York: Columbia University Press, 2013).

3. For an excellent treatment of this theme, see Mick Smith, *Against Ecological Sovereignty: Ethics, Biopolitics and Saving the Natural World* (Minneapolis: University of Minnesota Press, 2011).

4. Ibid.

5. My evangelican Christian neighbor took issue with my *Treehuggers* bumper sticker: "Nature is my bitch, I will do with her what I will."

6. Bruno Latour, *Facing Gaia. Eight lectures on the New Climatic Regime* (Oxford: Polity, 2015).

7. Martin Heidegger, "Letter on Humanism," *Martin Heidegger: Basic Writings* (London: Routledge and Kegan Paul, 1976), 206.

8. Giorgio Agamben, *The Open: Man and Animal* (Stanford: Stanford University Press, 2003).

9. J. M. Coetzee, *The Lives of Animals* (Princeton: Princeton University Press, 2016).

10. Charles Patterson, *Eternal Treblinka: Our Treatment of Animals and the Holocaust* (Seattle: Lantern, 2002).

11. Friedrich Nietzsche, *Genealogy of Morals* (Oxford: Oxford University Press, 2009).

12. See Mark Bekoff, *The Emotional Lives of Animals* (New York: New American Library, 2008); Donna Haraway, *When Species Meet* (Minneapolis: University of Minnesota Press, 2008); Vicki Hearne, *Adam's Task* (New York: Skyhorse, 2016); and Jane Goodall's works on chimps.

13. See the Einstein-Freud correspondence (1931–32) on war.

14. "The Holocene extinction, sometimes called the Sixth Extinction, describes the ongoing extinction event of species during the present Holocene epoch (since around 10,000 BCE) mainly due to human activity. The large number of extinctions span numerous families of plants and animals including mammals, birds, amphibians, reptiles and arthropods. . . . The vast majority are undocumented. . . . The present rate of extinction may be up to 140,000 species per year." (Wikipedia, modified]

15. The Human Microbiome Project, launched in 2008 is an NIH initiative (USA) aimed at identifying the microorganisms found in human bodies (the human microbiome).

16. See Michael Schellenberger and Ted Nordhaus, *Break Through: From the Death of Environmentalism to the Politics of Possibility* (New York: Houghton Mifflin, 2007), in which they castigate environmentalists for not offering a positive vision.

17. The concrete End of Man! Whatever else happens in the meantime, here is an extract from a narrative about our very distant fate: "About 1.2 billion years from now, the sun will begin to change. As the hydrogen fuel in its core is used up, the burning will spread outward toward the surface. This will make the sun grow brighter. This increased radiation will have a devastating effect on our planet. . . . The mean surface temperature of the earth will rise from about 68°F to 167°F. The earth's oceans will evaporate. The planet will become a stark, lifeless desert. . . . At the age of about 11–12 billion years . . . the mountains of the earth will melt and flow like red-hot molasses into vast, flat seas of lava. A bloated red sun will fill more than half the sky."

All is not gloom however: "While this spells the death of the inner planets, it will bring new life to the more distant worlds." Ron Miller, "What the Death of the Sun Will Look Like," io.gizmodo.com, April 9 2013.

3. COSMIC PASSIONS

1. David Hume, *A Treatise of Human Nature*, ed. L. A. Selby-Bigge (Oxford: Clarendon Press, 1975). I do not offer here a serious scholarly engagement with Hume's thinking about the passions. His classic claim operates with binaries (reason and passion), and his currency is pain and pleasure. Those allergic to this language will treat it with caution. What we call pain and pleasure are always more specifically manifested and within frames of significance. Importantly, too, Hume distinguishes direct and indirect passions, depending on how immediately they connect with the pain or pleasure afforded by an experience, and whether our sense of self is involved. To the extent that his despair and my angst overlap, we disagree about how to classify this passion. For Hume, it is a direct passion. I will try to think it alongside Kant's sublime, which itself would seem a development of Hume's indirectness. As for Hume's joy and my delight, I will emphasize the second order ramifications of delight in the natural world: a visceral investment in our connectedness.

2. This is the space from which the "witty and attractive Thracian servant-girl is said to have mocked Thales for falling into a well while he was observing the stars and gazing upwards." If stargazing Thales subsequently predicted a solar eclipse, and made a fortune betting on olive press futures, who can doubt that as a geologically alive being, he was equally susceptible to the passions we might call cosmic awe and connectedness?

3. G. W. Leibniz, "The Principles of Nature and of Grace, Based on Reason," in *Leibniz Selections*, ed. Philip P. Wiener (New York: Charles Scribner's Sons, 1951), 527.

4. Plato, *Theaetetus* (Oxford: Oxford University Press, 2014), 155d.

5. Rene Descartes, *The Passions of the Soul*, Art 53 in *The Passions of the Soul and Other Late Philosophical Writings* (Oxford: Oxford University Press, 2016).

6. See also Leibniz, "On the Ultimate Origin of Things (1697).

7. Brian Swimme, *The Universe Story: From the Primordial Flaring Forth to the Ecozoic Era—A Celebration of the Unfolding of the Cosmos* (New York: Harper, 1994).

8. See Friedrich Nietzsche, "The *Three Metamorphoses*" in *Twilight of the Idols* (Harmondsworth: Penguin, 1990).

9. See Luce Irigaray "Sexual Difference" in *An Ethics of Sexual Difference* (Ithaca, NY: Cornell University Press, 1993).

10. This, I believe, is Heidegger's position. See, for example, his riff on *deinon* in Sophocles's *Antigone* and his further treatment of the *unheimliche* (uncanny) in *An Introduction to Metaphysics* (Garden City, NY: Anchor, 1961), 127. See also the excellent extended treatment in Katherine Withy, *Heidegger on Being Uncanny* (Cambridge, MA: Harvard University Press, 2015).

11. Curiosity (*Neugier*), along with ambiguity (*Zweideutigkeit*) and idle talk (*Gerede*). See Martin Heidegger, *Being and Time* (Albany: SUNY Press, 2010), sections 36, 37.

12. Cannibals, torturers, and psychopaths may enjoy their own forms of curiosity and delight.

13. Immanuel Kant, *Critique of Judgment* (Oxford: Oxford University Press, 2009), Section 28.

14. Martin Heidegger, "*What Is Metaphysics?*" in *Martin Heidegger: Basic Writings* (London: Routledge and Kegan Paul, 1978).

15. Friedrich Nietzsche, *Thus Spoke Zarathustra* (Harmondsworth: Penguin, 1961).

16. Dale Jamieson, *Reason in a Dark Time* (Oxford: Oxford University Press, 2014), is nodded to in the subtitle of this book, along with Hannah Arendt's *Men in Dark Times*. Arendt in turn was alluding to Brecht, who had written: "In the dark times / Will there also be singing? / Yes, there will also be singing / About the dark times." Motto to "Svendborg Poems," in *Poems 1913–1956* (London: Routledge, Chapman and Hall, 1939).

17. This is a regular theme in the climate-change literature. See, for example, John M. Meyer, *Engaging the Everyday: Environmental Social Criticism and the Resonance Dilemma* (Cambridge, MA: MIT Press, 2015), which explores the need for some sort of resonance in everyday life with otherwise distant environmental issues for action to take place.

18. Reportedly adopted by students at the University of East Anglia, UK, October 2014.

19. See Friedrich Nietzsche, *The Genealogy of Morals* (Oxford: Oxford University Press, 2009). *Ressentiment* is not just a feeling, not even a personal passion, but a pervasive mood infecting a culture, with the deepest ramifications.

20. Happiness studies is a growth industry. There is even a *Journal of Happiness Studies*.

4. THINKING GEOLOGICALLY AFTER NIETZSCHE

1. These questions are not new. Derrida raises such issues in his discussion of the responsibilities of inheritance—what we can learn from Marx even after we have buried him. See *Specters of Marx: The State of the Debt, The Work of Mourning & the New International* (London: Routledge, 2006). The temptation

of pessimism is closely allied to the confrontation with nihilism central to Nietzsche's thinking, and to Heidegger's treatment of it. Kant mused on what was even then an old question: "Is the human race constantly progressing?" If we conclude that this is not so, it is not difficult for nihilism to fill the vacuum. But, again, things may not be quite so simple.

2. Friedrich Nietzsche, "Uses and Disadvantages of History for Life," in *Untimely Meditations* (Cambridge: Cambridge University Press, 1983), 76.

3. When one considers that both individuals and nations can be thought of as multiplicities struggling to reconcile warring factions, the analogy between the two becomes more plausible.

4. Nietzsche, *Untimely Meditations*, 77.

5. Ibid.

6. Immanuel Kant, *Universal Natural History and Theory of the Heavens* (1755) (New York: Richer Resources Publications, 2009).

7. The term Anthropocene was introduced in 2000 by Eugene Stoermer and Paul Crutzen.

8. Roxanne Dunbar-Ortiz, *An Indigenous Peoples' History of the United States (Revisioning American History)* (Boston: Beacon Press, 2015).

9. See Indian Removal Act (1830). "By 1837, 46,000 Native Americans from these southeastern states had been removed from their homelands, thereby opening 25 million acres for predominantly white settlement." (Wikipedia) This involved five nations, including the Choctaw, Seminole, and Cherokee. "Redskins" was the word used for the bodies of Indians killed for bounty scalps from 1835 onward. To prevent fraud (cutting a scalp in half), two ears were required.

10. Editorial review by Robert S. Desowitz in *Scientific American* (2005).

11. This is the logic behind the need for coercion in Garrett Hardin's analysis of the tragedy of the commons, first published in 1968 and often reprinted. See his "The Tragedy of the Commons," *Science* 162 (3859): 1243–1248.

12. I am speaking of full *philosophical* consideration. It is clear that race, gender, and class inequality are everywhere alive and kicking.

13. While I have been teaching philosophy, half the mammals on the planet have died out.

14. Nietzsche, *Untimely Meditations*, 76.

5. ANGST AND ATTUNEMENT

1. Martin Heidegger, *Being and Time* (Albany: SUNY Press, 2010), Section 74.

2. Ibid., Sections 56, 57.

3. Glenn Greenwald interviewed on *Democracy Now*, May 26 2017.

4. The great British physicist Isaac Newton's Third Law of motion says, "Every action has an equal and opposite reaction" (1687). No one accused him of being unpatriotic. But the idea that we might be a contributing cause to our own suffering is seen in that way.

5. For a groundbreaking Deleuze-inspired discussion of the politics of passion, see John Protevi, *Political Affect: Connecting the Social and the Somatic* (Minneapolis: University of Minnesota, 2009).

6. "Haunted by the Future," *Research in Phenomenology* 36 (2006): 274–295.

7. Bertrand Russell says that farmyard chickens will conclude inductively after many weeks of corn from the farmer that all is well. One day he wrings their necks. So much for induction. *Problems of Philosophy* (1910) (Oxford: Oxford University Press, 1997), Chapter VI (On Induction).

8. Martin Heidegger, *The Fundamental Concepts of Metaphysics: World, Finitude, Solitude* (Bloomington: Indiana University Press, 2001).

9. Søren Kierkegaard, *Either/Or* (Harmondsworth: Penguin, 1992).

10. See note 3.

11. Jacques Derrida and Giovanna Borradori, *Philosophy in a Time of Terror* (Chicago: University of Chicago Press, 2004).

12. Rainer Maria Rilke, "Archaic Torso of Apollo," in *The Selected Poetry of Rainer Maria Rilke*, trans. Stephen Mitchell (New York: Vintage, 1989).

13. Jacques Derrida, *Writing and Difference* (London: Routledge, 2001).

14. See J. M. Coetzee, *Elizabeth Costello* (Harmondsworth: Penguin, 2004).

15. Immanuel Kant, *Idea for a Universal History from a Cosmopolitan Point of View* (1784), trans. Lewis White Beck, in *On History* (Indianapolis: Bobbs-Merrill, 1963). See Jacques Derrida, "An Apocalyptic Tone Recently Adopted in Philosophy" which skewers Kant's rationalism. In *Raising the Tone of Philosophy, Late Essays by Immanuel Kant, Transformative Critique by Jacques Derrida*, ed. Peter Fenves (Baltimore: Johns Hopkins University Press, 1993).

16. Sam Scheffler, *Death and the Afterlife* (Oxford: Oxford University Press, 2016).

17. Heidegger, *Fundamental Concepts of Metaphysics*, Section 18.

18. Martin Heidegger, *Contributions to Philosophy*, trans. Richard Rojcewicz and Daniela Vallega-Neu (Bloomington: Indiana University Press, 2012), esp. 5.

19. This is what is needed, according to Garrett Hardin's tragedy of the commons theory (1968).

6. THE PRESENT AGE: A CASE STUDY

1. Immanuel Kant, *Idea for a Universal History with a Cosmopolitan Aim* (Cambridge: Cambridge University Press, 2009).

2. Søren Kierkegaard, *The Present Age* (New York: Harper, 1962).

3. Friedrich Nietzsche, *The Genealogy of Morals* (New York: Vintage, 1989).

4. Martin Heidegger, *Contributions to Philosophy*, trans. Richard Rojcewicz and Daniela Vallega-Neu (Bloomington: Indiana University Press, 2012), #6.

5. Martin Heidegger, *An Introduction to Metaphysics* (New York: Anchor, 1961), 38.

6. In Jacques Derrida, *Specters of Marx* (New York: Routledge, 2012).

7. Here are a selection of twenty from the web (summer 2016)—deliberately unattributed:

1. "We've got a lot of reasons to be angry. But the country is in a very Dark-Side-of-the-Force mood, convinced that anger is empowering, not blinding."

2. "The article . . . tells a very different story—rising despair, despondency, disengagement, and in some cases delusional behaviour, in the wake of serial climate disappointments."

3. "Market jitters: Anxiety about the state of the market given the development of certain unfavorable market conditions."

4. "Yet no matter what the nomenclature (some refer to the problem as 'ecoanxiety,' while others talk about 'doomer depression' and 'apocalypse fatigue'), despondency over what many believe is societal failure to adequately acknowledge or address environmental issues has become a line of psychological inquiry."

5. "The E.U.'s failures have produced a frightening rise in reactionary, racist nationalism—but Brexit would, all too probably, empower those forces even more."

6. "People are angry . . . they feel nothing is being done."

7. "I do not underestimate Trump's appeal. He's harnessed the national anger."

8. "In the working-class east London borough of Havering, there was anger mixed with hope Thursday, as locals voted in Britain's landmark EU referendum. Citing exasperation at uncontrolled migration and a fierce desire for independence from Brussels, a string of voters said they wanted Britain out of the EU for good."

9. "The mood in Brussels grew ever darker on Wednesday, with EU leaders taking a hard line over the terms of the U.K.'s exit."

10. "Without hope, the horror of climate change paralyzes rather than politicizes."

11. "News about climate change is almost always alarming, depressing, or both. But Tim Flannery believes there is qualified hope that things may get better."

12. "If I didn't know he was campaigning for the presidency, I'd assume Trump's entire goal was to foment anger and hatred."

13. "The leave vote has led to 'a kind of hysteria, a contagious mourning' among part of the population."

14. "They also tend to be the kinds of voters who are most skeptical of economic globalization, and the prominent role that China [is] playing in the

current market jitters may serve to underscore those feelings. If that's the case, the effects of Monday's shock may fade, but the anxiety it plays into figures to linger."

15. "For others who have struggled to find meaning in their lives, ISIS is a thrilling cause and call to action that promises glory and esteem in the eyes of friends, and eternal respect and remembrance in the wider world that many of them will never live to enjoy."

16. "For at least some of the Islamic State's young volunteers there is a feeling of joy and celebration involved in joining up."

17. "Administration officials described "terrorists" as hateful, treacherous, barbarous, mad, twisted, perverted, without faith, parasitical, inhuman, and, evil. Americans, in contrast, were described as brave, loving, generous, strong, resourceful, heroic, and respectful of human rights."

18. "Faludi finds that even in the world they supposedly own and run, men are at the mercy of cultural forces that disfigure their lives and destroy their chance at happiness. As traditional masculinity continues to collapse, the once-valued male attributes of craft, loyalty, and social utility are no longer honored, much less rewarded."

19. "The author turns her attention to the masculinity crisis plaguing our culture at the end of the '90s, an era of massive layoffs, 'Angry White Male' politics, and Million Man marches."

20. "The big takeaway," Krugman insists, is that "climate despair is all wrong" and another example of the supposedly serious people in America embracing a destructive, but emotionally appealing, policy."

8. See Martin Heidegger, *The Fundamental Concepts of Metaphysics: World, Finitude, Solitude* (Bloomington: Indiana University Press, 2001).

9. Heidegger, *Introduction to Metaphysics*, 1.

10. Heidegger, *Contributions to Philosophy*.

11. For justification of continuing states of emergency and so on, see Carl Schmidt, *The Concept of the Political* (Chicago: University of Chicago Press, 2007). See also Chapter 10.

12. This is the plausible verdict of the film *The Corporation* (2003).

13. The classic source here is Karl Marx on "Estranged Labor," in *The Economic and Philosophical Manuscripts of 1844* (New York: International, 1964).

14. Heidegger, *Fundamental Concepts of Metaphysics*, 77.

15. See the Introduction to my *The Step Back: Ethics and Politics after Deconstruction* (Albany: SUNY Press, 2005).

7. POSTHUMANIST RESPONSIBILITY

1. This is part of the point of Derrida's famous essay on Levinas, "Violence and Metaphysics," in *Writing and Difference* (London: Routledge, 2001).

2. See Francois Raffoul, *The Origins of Responsibility* (Bloomington: Indiana University Press, 2010). He draws on Levinas, Heidegger, and Derrida.

3. Theodor Adorno, *The Jargon of Authenticity* (London: Routledge, 2002).

4. Campaigners in the Netherlands took their government to court for failing to protect its citizens from climate change. They want to compel the Dutch government to reduce its carbon emissions to 40 percent below 1990s levels by 2020 and to declare that global warming of more than 2 degrees C will violate fundamental human rights. In June 2005, the court in The Hague ordered the Dutch government to cut its emissions by at least 25 percent within five years.

5. Newsflash: "New York sues big oil companies over climate change," *Financial Times*, January 10, 2018. "Some of the world's largest listed energy companies are facing a lawsuit for 'billions of dollars' after New York City accused them of contributing to climate change."

6. The CIA coined this phrase in the 1960s: withholding information from senior officials to protect them from liability.

7. See David Ray Griffin, *Unprecedented: Can Civilization Survive the CO2 Crisis?* (Oxford: Clarity Press, 2015).

8. It is important to note that when Heidegger says that "science does not think," this is not an argument against science, and gives no grounds for climate science skepticism, or claims that evolution is "just a theory."

9. This would be appropriate for bump fire stock manufacturers. These aftermarket devices convert rifles into automatic weapons, enabling Las Vegas–type carnage (2017).

10. A good recent example of this is Griffin's *Unprecedented*, which unrelentingly exposes the machinations of Big Oil in constructing the "debate" about climate change.

11. This should have been the Stern Review (2006) by a British economist arguing that precautionary eco investment now would reap substantial benefits in the future.

12. See Pew Research Center, *The Politics of Climate*, October 4, 2016.

13. See Jacques Derrida, *The Gift of Death* (Chicago: University of Chicago Press, 2007).

14. Martin Heidegger, *Contributions to Philosophy (Of the Event)* (Bloomington: Indiana University Press, 2012).

15. Jacques Derrida and Anne Dufourmantelle, *Of Hospitality* (Stanford: Stanford University Press, 2000).

16. See his *De Rerum Natura* and his account of the clinamen.

17. Gerard Manley Hopkins borrowed this term from Duns Scotus and gave it new prominence. It means the specific "thisness" of something.

18. See Jacques Derrida, *The Death Penalty*, vol. 1, trans. Peggy Kamuf (Chicago: University of Chicago Press, 2013).

8. THE NEW MATERIALISM

1. New materialism is represented in the work of people such Karen Barad, Jane Bennett, Diana Coole, Bruno Latour, Samantha Frost, Brian Massumi, Rosi Braidotti, and Manuel DeLanda.

2. Friedrich Nietzsche, "Truth and Falsity in their Ultra-Moral Sense," in *The Existentialist Tradition*, ed. Nino Languilli (Atlantic Highlands, NJ: Humanities Press, 1971).

3. Søren Kierkegaard, *Fear and Trembling* (Harmondsworth: Penguin, 1986).

4. From his "Immanence and Transcendence in the Genesis of Form," in *A Deleuzian Century*, ed. Ian Buchanan (Durham, NC: Duke University Press, 1999), 129.

5. See Timothy Morton, *Ecology without Nature: Rethinking Environmental Aesthetics* (Cambridge, MA: Harvard University Press, 2009); *The Ecological Thought* (Cambridge, MA: Harvard University Press, 2012); and *Hyperobjects: Philosophy and Ecology after the End of the World* (Minneapolis: University of Minnesota Press, 2013).

6. See Joe Hughes on Deleuze and joy, in *Philosophy after Deleuze* (London: Bloomsbury, 2012), 76–77.

7. These remarks also bear on Heidegger's attempts in *Contributions to Philosophy* to move away from thinking "about" to a more inceptual thinking.

8. See books by Karen Barad, *Meeting the Universe Half Way* (Durham, NC: Duke University Press, 2007); Jane Bennett, *Vibrant Matter: A Political Ecology of Things* (Durham, NC: Duke University Press, 2010); *New Materialisms: Ontology, Agency and Politics*, ed. Diana Coole and Samantha Frost (Durham, NC: Duke University Press, 2010); *New Materialism: Interviews and Cartographies*, ed. Rick Dolphijn and Iris van der Tuin (Ann Arbor: University of Michigan, Open Humanities, 2012). In the background throughout are Deleuze, Whitehead, and Spinoza. See also Nancy Tuana, "Being Affected by Climate Change: The Anthropocene and the Body of Ethics," in *Ethics in the Anthropocene*, ed. Kenneth Shockley and Andrew Light (Cambridge, MA: MIT Press, 217).

9. "Matter feels, converses, suffers, desires, yearns and remembers." Interview with Karen Barad, in *New Materialism*, 3.

10. Ibid.

11. See Tuana, "Being Affected by Climate Change."

12. My own substantial discussion of Whitehead's importance for environmental thinking can be found in *Reoccupy the Earth* (New York: Fordham

University Press, 2018), Chapter 3, "Ecological Imagination: A Whiteheadian Exercise in Temporal Phronesis."

13. See writings by Naomi Klein, Gus Speth, Herman Daly, and many others.

14. This verdict will not be seen as a criticism by those (like Rorty) who believe that is all philosophy can ever do (discursive innovation, new ways of talking). New materialism, however, thinks it is doing more than that.

15. Specifically, I would draw attention to *Deleuze Studies* 10, no. 4 (November 2016): "Deleuze and Guattari in the Anthropocene."

16. These include Rosi Braidotti, Claire Colebrook, James Williams, Keith-Ansell Pearson, Tamsin Lorraine, Paul Patton, Brian Massumi, Ronald Bogue, John Protevi, Joe Hughes, Constantin V. Boundas—just to begin a long list.

17. Gilles Deleuze and Felix Guattari, *What Is Philosophy?* (New York: Columbia University Press, 1994), 99–100.

18. Ibid., 108. They continue: "Europeanization does not constitute a becoming but merely the history of capitalism, which prevents the becoming of subjected peoples. Art and philosophy converge at this point: the constitution of an earth and a people that are lacking as the correlate of creation. This people and earth will not be found in our democracies. Democracies are majorities, but a becoming is by its nature that which always eludes the majority."

19. See Rodolphe Gasché, *Geophilosophy* (Evanston, IL: Northwestern University Press, 2014).

20. Chapter 4 of Deleuze and Guattari, *What Is Philosophy?*

21. Gilles Deleuze and Claire Parnet, *Dialogues II* (London: Athlone, 1987), 40.

22. See Naomi Klein, *This Changes Everything: Capitalism vs. the Climate* (New York: Simon & Shuster, 2015).

23. Deleuze and Guattari, *What Is Philosophy?* 110.

24. Ibid., 109.

25. Such as Nick Land. See his *Teleoplexy: Notes on Acceleration* (London: Urbanomic Press, 2014).

26. Rosi Braidotti, "Affirming the Affirmative: On Nomadic Affectivity," *Rhizomes* 11/12 (Fall 2005–Spring 2006), section 16. For a fuller story, see her *The Posthuman* (Cambridge: Polity, 2013).

9. THE UNTHINKABLE AND THE IMPOSSIBLE

1. James Hansen wrote this with regard to the Canadian Keystone pipeline in the *New York Times* (May 9, 2012):

If Canada proceeds, and we do nothing, it will be game over for the climate.
Canada's tar sands, deposits of sand saturated with bitumen, contain twice the amount of carbon dioxide emitted by global oil use in our entire history. If we were to fully exploit this new oil source, and continue to burn our conventional oil, gas and coal supplies, concentrations of carbon dioxide in the atmosphere eventually would reach levels higher than in the Pliocene era, more than 2.5 million years ago, when sea level was at least 50 feet higher than it is now. That level of heat-trapping gases would assure that the disintegration of the ice sheets would accelerate out of control. Sea levels would rise and destroy coastal cities. Global temperatures would become intolerable. Twenty to 50 percent of the planet's species would be driven to extinction. Civilization would be at risk.

2. If the Arctic sea ice melts, large currents such as the Atlantic Conveyor could slow or even reverse. Without the vast heat from these ocean currents, Europe's average temperature would drop 5 to 10° C, with East Coast America chilled a little less. This would take us back roughly to the last ice age.

3. See Vincent Sarich and Frank Miele, *Race: The Reality of Human Differences* (Boulder: Westview, 2004), on preparations for a genetic bomb. They argue that the Human Genome Project could be used to selectively target racially distinct humans.

4. The deeply disturbing history of forced sterilization in the US and elsewhere is summarized at http://abcnews.go.com/onair/2020/2020_000322_eugenics_feature.html.

5. This sense of a stark choice awaiting us is increasingly shared. See, for example, Geoff Mann and Joel Wainwright, *Climate Leviathan: A Political Theory of Our Planetary Future* (London: Verso, 2018).

6. William McDonough and Michael Braungart, *Cradle to Cradle: Remaking the Way We Make Things* (Berkeley, CA: North Point Press, 2002).

7. See Project for a New American Century, *Rebuilding America's Defenses* (1997).

8. See Naomi Klein, *The Shock Doctrine: The Rise of Disaster Capitalism* (London: Picador, 2008).

9. This diagnosis is not new: "We are the hollow men/We are the stuffed men/Leaning together/Headpiece filled with straw. Alas!/Our dried voices, when/We whisper together/Are quiet and meaningless." T. S. Eliot, "The Hollow Men" (1925).

10. See the excellent short account in Barbara Muraca's *Living Well: A Society Beyond Growth* (Albany: SUNY Press, 2019).

11. Robert Pogue Harrison, *Forests: The Shadow of Civilization* (Chicago: University of Chicago Press, 1993), 249.

12. The need to act right now or not at all arguably can impose a competing responsibility. Avoiding rigid rule-following may not always be possible,
let alone desirable. Think of swerving to avoid a stray dog in the road.

10. WHAT IS TO BE DONE? DEMOCRACY AND BEYOND

1. See Dave Foreman, *Ecodefense: A Strategic Guide to Monkeywrenching*
(Tucson, AZ: Ned Ludd Books, 1987), 10–17.

2. I do not for a moment deny that green consumerism is often ineffectual
alibi-generating hype.

3. Paul Mason, *Postcapitalism: A Guide to Our Future* (New York: Farrar,
Straus and Giroux, 2016).

4. See, for example, Oliver Morton, *The Planet Remade: How Geoengineering Could Change the World* (Princeton: Princeton University Press, 2015).

5. One of the best is Naomi Oreskes and Eric M. Conway, *Merchants of
Doubt: How a Handful of Scientists Obscured the Truth on Issues from Tobacco
Smoke to Global Warming* (London: Bloombsury, 2011).

6. For an excellent survey of what may be in store for us, see David
Wallace-Wells, "The Uninhabitable Earth Annotated Edition," *New York
Magazine*, July 14, 2017.

7. Joseph Romm, *Climate Change: What Everyone Needs to Know* (Oxford:
Oxford University Press, 2016).

8. Oreskes and Conway, *Merchants of Doubt*.

9. Andrew J. Hoffman, *How Culture Shapes the Climate Change Debate*
(Stanford: Stanford University Press, 2015); Kari Marie Norgaard, *Living in
Denial: Climate Change, Emotions, and Everyday Life* (Cambridge, MA: MIT
Press, 2011).

10. Paul Hawken, ed., *Drawdown: The Most Comprehensive Plan Ever
Proposed to Reverse Global Warming* (New York: Penguin, 2017; Michael
Bloomberg and Carl Pope, *Climate of Hope: How Cities, Businesses, and Citizens
Can Save the Planet* (New York: St. Martin's, 2017); Bill McKibben, *Eaarth:
Making a Life on a Tough New Planet* (New York: St. Martin's, 2011).

11. Naomi Oreskes and Erik M. Conway, *The Collapse of Western Civilization: A View from the Future* (New York: Columbia University Press, 2014).

12. Stephen M. Gardiner, *A Perfect Moral Storm: The Ethical Tragedy of
Climate Change* (New York: Oxford University Press, 2011); Dale Jamieson,
*Reason in a Dark Time: Why the Struggle Against Climate Change Failed—and
What It Means for Our Future* (New York: Oxford University Press, 2014).

13. Naomi Klein, *This Changes Everything: Capitalism vs. the Climate* (New
York: Simon & Shuster, 2015); David Ray Griffin, *Unprecedented: Can Civilization Survive the CO2 Crisis?* (Atlanta, GA: Clarity, 2015). Griffin's book has a
comprehensive brief.

14. Sallie McFague, *A New Climate for Theology: God, the World, and Global Warming* (Westminster, KY: Fortress, 2008); Pope Francis, *Laudato Si—On Care for Our Common Home* (New York: Penguin, 2015).

15. The significance of this thought is explored in my "Toxicity and Transcendence: Two Faces of the Human," in *Thinking Plant Animal Man* (Minneapolis: University of Minnesota Press, 2019).

Adorno, Theodor, 63, 85, 150n3
affect, 60
the affective, 2, 6, 25, 43, 103, 126
affective resonance, 56
Agamben, Giorgio, 30, 143n8
agency, 1, 9, 24, 78, 84, 86, 89, 91, 100,
    109, 118, 120; and angst, 42; aporetics
    of, 65, 66–67; autonomous, 87;
    geologically informed, 95; human, 94;
    individual, 43; instrumentalist
    understanding of, 125; materialist
    understanding of, 21, 90–91, 96,
    101–102, 103; and mood, 74; shared,
    10, 25, 82, 96, 103, 131; "we" of,
    71–72
alienation, 77, 81, 117, 134
ambiguity, 35, 145n11
angst, 24, 37, 44, 62–63, 73, 144n1;
    cosmic, 42, 62; geological, 42, 43, 52,
    59; Heidegger on, 41, 43, 59, 62–63;
    individual, 75; modern, 18; and the
    sublime, 40–42
the Anthropocene, 5, 6, 51, 55, 102,
    139n2, 146n7; Age, 14; and climate
    change, 2, 17, 29, 61; and conscious-
    ness, 29, 60; and Deep Time, 24; as
    event, 22, 70; and geological angst, 42;
    and Heidegger, 8, 69; and history, 10;
    and hospitality, 24; and Nietzsche, 1;
    and the Sixth Extinction, 32–33, 42;
anthropocentrism, 16, 20, 30, 133
anthropological machine, 30

apocalyptic scenarios, 70, 107
Ardipithecus ramidus, 5
Arendt, Hannah, 7; Men in Dark Times,
    145n16
Australopithecus afarensis, 27
autonomy, 23, 85, 131

Barad, Karen, 101–102, 151n1
Beckett, Samuel, 52, 70
Bekoff, Mark, 32
Benford, Gregory, 141n30
Bennett, Jane, 102, 151n1; Vibrant
    Matter, 77, 101
Bentham, Jeremy, 32
Berry, Thomas: The Great Work, 77
Berry, Wendell, 117, 130; The Unsettling
    of America, 117
Bess, Michael: Our Grandchildren
    Redesigned, 77
biopolitics, 110, 116
Blanchot, Maurice, 21, 70
boredom, 61, 72, 76–77, 81
Braidotti, Rosi, 103, 106, 151n1,
    152n16
Brundtland Report, 83, 114
Burke, Edmund, 1, 9

capitalism, 105, 139n2, 152n18; carbon,
    43; consumer, 105–106; free market,
    69, 102, 119
Chomsky, Noam, 121, 123
Churchill, Winston, 9, 121

climate change, 2, 37, 53, 70, 74, 107,
    130, 133–134, 148nn10–11; anthropo-
    genic, 24, 133; apocalyptic character
    of, 54, 107; and business interests,
    85, 87, 150nn8,10; catastrophic, 17,
    38, 66, 75, 87, 88, 104, 108; challenge
    of, 45; global, 62, 65, 82, 85, 95,
    141n33; and Heidegger, 68, 75, 81;
    impact of, 35; irreversible, 17, 123,
    124, 127; literature on, 132, 145n17;
    and politics, 14, 72, 81, 108, 123, 131,
    150n4; prevention, 17, 108; and
    resignation, 2, 14, 60, 75, 111;
    resistance to, 38, 90; response to, 2,
    36, 44, 100; responsibility for, 86–87,
    102; and scarcity, 111; urgency
    of, 57, 61
climate crisis, 22, 42, 61, 66, 70, 103, 104,
    105, 109, 126, 131, 140n8
climate denial, 85, 86, 87, 88, 133,
    150n8
Coetzee, J. M., 31
Colebrook, Claire, 103, 106, 152n16
collective action, 43, 65, 82, 125, 128
conatus, 101
consciousness, 16, 22, 33, 90, 97;
    collective, 76; cosmic, 35; critical, 17,
    59; deep time, 1; geological, 29, 35, 44,
    58, 60, 90, 96–97, 120, 133; historical,
    19; human, 29, 33, 34; self-, 16, 93;
    species-, 35
constitutive relationality, 101
consumerism, 69, 85, 112, 114, 116–117,
    119, 139n2; environmentally friendly,
    125; green, 127, 154n2
Conway, Eric M., 142n2, 154n5,
Coole, Diane, 101, 151n1
Copernicus, Nicolaus, 6, 23, 28, 34
Costello, Elizabeth, 70
creative connectivity, 131, 135
Crick, Francis, 34
culpable ignorance, 88
culpable negligence, 86–88
curiosity, 37, 39, 40, 44, 45, 145nn11,12
cybercitizenship, 130

Darwin, Charles, 6, 23, 28–29, 34
Deep Time, 2, 16, 17, 23, 71, 122, 133;
    and affect, 126; and anthropocen-
    trism, 20–21, 31; burden of, 69;
    cosmic, 15; and democracy, 124;
    events, 15; and geological conscious-
    ness, 6, 29, 60–61, 119, 133; and
    Heidegger, 69; and hospitality, 23–24;
    knowledge of, 19; and nested time, 11;
    and the past, 5, 14; reservoirs, 22; Ted
    Toadvine on, 7, 15
DeLanda, Manuel, 96–98, 151n1
Deleuze, Gilles, 8, 17, 21, 96, 100,
    103–106, 134, 147n5, 151nn6,8;
    Nietzsche and Philosophy,103
delight, 37–38, 40, 44–45, 144n1, 145n12
democracy, 43, 72, 81, 121–124, 128;
    -to-come, 51, 100; representative,
    122–123, 125
Derrida, Jacques, 7, 53, 73, 74, 82, 93,
    100, 104, 107, 111, 118, 132, 149n1,
    150n2; on 9/11, 18, 41–42, 65, 69; *The
    Animal That I Am*, 22; on animals, 22,
    31, 32; the autoimmune, 53; decon-
    struction, 99, 100; and deep time, 2,
    23; on Enlightenment values, 51; on
    Heidegger, 21, 30, 32; on hospitality,
    23, 90; on the "im-possible," 66, 70,
    89, 111, 118; on Nancy, 21; on
    responsibility, 88–90, 145n1; *Specters
    of Marx*, 100; and time, 8; on the
    to-come, 51, 61, 70, 100, 134; on the
    undecidable, 2, 58, 118
Descartes, René, 36, 37, 93, 144n5
desire, 51, 69, 85, 92–93, 140n8, 151n9;
    and aporetic time, 19; and autonomy,
    85; and consumerism, 45, 52, 104, 105,
    112, 114, 116, 130, 134; for control, 30,
    69, 118; Deleuze on, 103–105, 134;
    economy of, 113; false, 45, 116; and
    identity, 113, 116, 120, 130–131, 134
despair, 1, 33, 36–37, 42–45, 62, 73, 75
deterritorialization, 105
Diamond, Jared, 54; *Collapse*, 54
Dickens, Charles, 73

Dunbar-Ortiz, Roxanne, 52
dwelling, 1, 11–12, 44, 80, 93, 111, 132; authentic, 38; temporal, 19

empathic imagination, 83
the End of Man, 34, 143n17
the Enlightenment, 6, 34, 99; narrative, 29; project, 110; values, 50–51
the environment; damage to, 52, 80, 86; impact on, 7, 52, 87, 125; the issue of, 103, 145n17, 148n4; neglect of, 57; protection of, 125
Environmental Protection Agency, 126
environmentalism, 14, 89, 104, 128, 141n25, 143n16
the ethical, 69, 84, 94, 102, 106, 133; turn, 79
ethics, 82, 92, 93, 111, 123, 133; geological, 103; land, 4, 117; virtue, 117
evolution, 3–5, 19, 24, 30, 53, 133, 150n8; and biology, 32; cosmic, 16; and development, 6, 47, 121; and heritage, 32

Fleming, Pat, 47
forgetfulness of Being, 8, 57
Foucault, Michel, 4, 8, 19, 26, 29, 97, 100
freedom, 24, 84, 90–91, 103, 116; and ethics, 93; of expression, 86; individual, 114; personal, 53
Freud, Sigmund, 23, 27, 28, 31, 34, 57, 116, 143n13
Frost, Samantha, 151n1
fundamental attunement, 61, 72, 76–77
fundamental disposition, 68–69

Gaia, 38, 83; centrism, 30; hypothesis, 30
Gasché, Rodolphe, 103, 104
Gee, Henry: *In Search of Deep Time*, 8
geologically human, 34, 47
geology, 3, 39, 97
geophenomenology, 37
geophilosophy, 103–104, 152n19
globalization, 35, 81, 148n14
Goldmann, Lucian, 61

Goodall, Jane, 32, 143n12
Gore, Al: *An Inconvenient Truth*, 2, 43
grand narrative, 27, 105
Greenwald, Glenn, 146n3
Guattari, Felix, 7, 103–106
Guevara, Che, 66, 107

Habermas, Jürgen, 51
haecceity, 93
happiness, 45, 85, 112, 116, 149n18; studies, 45, 145n20
Haraway, Donna, 32
Harrison, Robert Pogue, 118
Hearne, Vicki, 32
Hegel, G. W. F., 5, 7, 9, 10, 21, 22, 68, 89, 92, 107
Heidegger, Martin, 2, 7, 21, 30, 59, 69, 73, 75, 79, 81, 85, 116, 132, 141n23, 145nn10,1, 146n1, 150n2; on animals, 32; on attunement, 71, 72, 76; *Being and Time*, 61, 67, 74; being towards death, 62, 70, 93, 118; on boredom, 76–77; *Contributions to Philosophy*, 67, 68, 72–75, 89, 118–119, 151n7; on curiosity, 39; Dasein, 6, 21–22, 67, 76, 92; dwelling, 80, 93; *Fundamental Concepts of Metaphysics*, 72, 76; on history, 6, 8, 22, 74; on humanism, 21; idle talk, 39, 145n11; "Language," 23; *Machenschaft*, 8, 43, 68; *das Man*, 63; on mood, 36, 43, 60, 61, 80; on ontology, 8, 57, 100; Other beginning, 8, 67–69, 74, 88; preparing the way, 67, 88, 89, 107, 118, 119; on responsibility, 89; on science, 20, 39, 95, 150n8; on the sun, 21–22, 24; on time, 8, 74, 76; *unheimlich* 44–45, 145n10
heteronomy, 23
historicity, 6, 22, 61, 140n6
historiography, 4, 6, 22, 140n6
history, 1, 2, 3, 12, 14, 23–24, 32, 47, 78, 80, 99, 107, 111, 124, 132; antiquarian, 48, 50–51, 59; apocalyptic, 6–7, 54, 108; the burden of, 1–2, 6, 9, 14, 25, 48, 60, 62, 67, 75, 77, 126; civilizational, 3,

history (cont.)
  5, 6; cosmic, 14, 133; critical, 48,
  50–53, 55, 56, 62; deep, 8, 58; end of,
  27; energetic, 4; evolutionary, 10, 27,
  29, 126; of forced sterilization, 153n4;
  geological, 1, 10, 29, 34, 46, 50, 52–53,
  61, 71, 96, 121, 133; German, 9;
  Heidegger on, 6, 74; human, 1, 3, 4, 5,
  9, 10, 12, 29, 32–34, 46, 49, 56, 58, 87,
  96, 128, 131; idea of, 9; of life, 29, 58,
  96; material, 22, 49, 68, 69, 118;
  monumental, 48, 50–51, 53, 55, 59, 99;
  naïve, 5; national, 9, 48, 49; natural, 5,
  58; Nietzsche on, 1, 9, 10,14, 46,
  47–49, 51, 58, 60–62, 67, 71, 75, 77,
  119, 126; of ontology, 57, 67, 100;
  planetary, 47, 121; posthumanist, 14; of
  repression, 31; species, 49; traumatic,
  9; Western, 5, 8, 49
Hobbes, Thomas, 55, 66, 94
Hodge, Joanna, 8
the Holocene 10, 15, 33, 143n14
Homo: erectus, 27; *habilis*, 27; *modesto*, 28;
  sapiens 5, 15, 26–27, 35, 51, 121, 133;
  sapiens neanderthalensis, 27; *sapiens
  sapiens*, 27
hospitality, 23, 24, 90
human exceptionalism, 31, 97
humanism, 20–21, 26, 84, 86; anti, 21;
  post, 21, 82, 86, 88, 96
Hume, David, 9, 36, 45, 64; on the
  passions, 35, 60, 144n1; on pleasure,
  45; on reason as the slave of the
  passions, 14, 25, 33, 36, 44
Husserl, Edmund, 7, 8, 20, 77, 91; *Crisis*,
  73; *Logical Investigations*, 91
hyperobject, 98

identity, 79, 116, 125, 128, 130; altruistic
  model of, 131; boundaries, 118;
  consumerist, 134; Deleuze and
  Guattari on, 104–106; and desire, 113,
  116, 120, 130; ethnic, 56; genetic, 114;
  investments, 79, 104; and memory, 31;
  species-, 34; value of, 104

the impossible, 18, 66, 107, 113, 116, 118,
  119, 126, 130; Blanchot on, 21; Derrida
  on, 66, 111, 118, 134; openness to, 85
  89, 134
individualism, 114, 115, 128; consumerist,
  113; possessive, 115
institutions, 27, 45, 85, 104, 134;
  cultural, 12; democratic, 122, 124;
  political, 18, 35, 126; social, 18, 95,
  126, 142
Irigaray, Luce, 7, 24, 37, 116

Jamieson, Dale: *Reason in a Dark Time*,
  43, 145n16
Jonas, Hans, 10

Kafka, Franz, 134
Kant, Immanuel, 7, 28, 49, 70, 73,
  141n28, 145n1; on agency, 66, 95; first
  antinomy, 16; on history, 5, 9; and
  responsibility, 86, 93; on self-
  consciousness, 93; on the sublime, 15,
  40–41, 43, 62, 144n1
Keller, Catherine: *Cloud of the Impossible*,
  77
Kierkegaard, Søren, 7, 62, 69, 73, 92,
  96–99, 100; *Concluding Unscientific
  Postscript*, 91; *Fear and Trembling*, 97;
  Sickness Unto Death, 62
Klein, Naomi, 70, 152n13; *This Changes
  Everything*, 7, 77
Krell, David, 140n12

Latour, Bruno, 30, 151n1
Leibniz, G. W., 37
Lemaître, Georges, 34
Leopold, Aldo, 4, 117
Levinas, Emmanuel, 79, 89, 92, 93,
  149n1, 150n2
Lorraine, Tamsin, 17, 152n16
Lukács, György: *History and Class
  Consciousness*, 61

Macy, Joanna, 47
Marcuse, Herbert, 7, 31

Marx, Karl, 7, 9, 23, 28, 68, 77, 81, 89, 107, 116, 145n1, 149n13
Mason, Paul, 127–128
Massumi, Brian, 60, 151n1, 152n16
materialism, 69, 91, 93; geologically informed, 94; new, 8, 21, 49, 82, 90, 96, 99, 100, 101, 102–103, 106, 151n1, 152n14; transcendental, 100, 104; vital, 77
Mazis, Glen, 7
McKibben, Bill, 109; *Do the Math*, 66
McPhee, John, 1; *Annals of a Former World*, 8
Meyer, John M., 145n17
microbiome project, 23, 33, 56, 72, 143n15
mood, 6, 14, 63, 73–75, 80, 81, 125, 126; Heidegger on, 36, 43, 61, 77; historical, 74; Nietzsche on, 61–62, 78, 145n19
Morris, David, 7
Morton, Timothy, 98–99, 100–101; *The Ecological Thought*, 98; *Ecology Without Nature*, 98; *Hyperobjects*, 98

Nagel, Thomas: *View from Nowhere*, 91
Nancy, Jean-luc, 21
the natural world, 27, 29, 30, 40, 56, 117, 118, 144n1
naturalism, 16, 28, 96; reductive, 103
nature, 11, 24, 28, 29, 30, 50, 97, 98, 99, 100, 142n5; deficit disorder, 56 118; hereditary, 48–49, 51; history of, 34; man's dominion over, 8, 30, 40, 68–69, 80, 103, 117; mechanistic, 103; second, 27, 48–49; and the sublime, 40; wonder of, 38
negative capability, 59, 81
negentropy, 4, 5, 16, 141n29
Never Again, 52, 99, 111
Nietzsche, Friedrich, 16, 22, 26, 37, 39, 43, 47, 55, 58, 71, 76, 89, 96–97, 116, 132, 134; active forgetting, 9, 47–48, 57–58, 73; on the burden of history, 1, 10, 14, 25, 48, 60, 62, 67, 75, 77, 126; and the death of God, 29; and deep

time, 20; eternal return, 23, 41; *The Genealogy of Morals*, 49; hangman's metaphysics, 65, 85; on history, 6, 8, 9, 46, 49, 51, 52, 61, 119, 133; and Kierkegaard, 97–100; on life, 6, 14, 48, 49, 51, 56; on morality, 31; perspectivism, 20; and responsibility, 48, 99, 126, 145n1; *ressentiment*, 44, 62, 73, 78, 145n19; on time, 8, 19; *Übermensch*, 49, 67; "The Uses and Disadvantages of History for Life," 1–2, 46, 47, 60, 67; Zarathustra, 43, 69
noosphere, 27, 130–131

ontology, 57, 67, 100, 103; process, 15; substance, 102

Pascal, Blaise, 41, 42, 62
the passions, 1, 36, 44, 47, 57, 75 78, 125, 126, 144n2; "animal," 31; and the Anthropocene, 2; curiosity, 39, 40, 44; and geophenomenology, 37; Hume on, 14, 25, 33, 35, 36, 44, 60, 144n1; Kierkegaard on, 96–97; Nietzsche on, 14, 25, 48; and reason, 45, 48, 55; wonder, 37, 40, 44
Patterson, Charles, 31
philosophy, 16, 22, 24, 57, 77, 106, 132, 152nn13,18; Greek, 5, 74
Plato, 14, 22, 37, 48, 57, 81, 122
Pleistocene, 15, 141n26
positionality, 94; irreducible, 91–92, 95
Protevi, John, 103, 152n16

realism, 13, 43, 90; agential 101
resignation, 1, 44, 54, 131; and the Anthropocene, 2, 24; reaction to climate change, 14, 43, 60, 61, 65–66, 75, 111; and *ressentiment*, 62, 63, 78
resoluteness, 62–63
responsibility, 24–25, 61, 71, 78, 82–83, 93, 114, 154n12; and capitalism, 139n2; and climate change, 82, 86, 102, 141n3; corporate, 72; and culpable negligence, 86–88; Derrida on, 100,

responsibility (cont.)
118–119, 145n1; ecological, 10;
environmental, 126; fiduciary, 86;
geological, 38–39, 43, 90, 130; and guilt,
42, 48, 85, 86, 88; and hospitality, 90;
human, 26, 34, 84; hyperbolic, 89, 90;
and materialism, 90, 96, 101; Nietzsche
on 48, 65; of the past for the present,
12–13; to the past, present, or future,
99, 100; and posthumanism, 96; and
resignation, 65; as response-ability, 85,
89–90; species, 10; terrestrial, 1, 83, 130
ressentiment, 44, 61–63, 73, 78, 80–81,
131, 145n19
revolution, 78, 89, 111, 124; economic,
127; French, 5, 50, 90; Industrial, 5,
43, 141n33; information, 128–129;
May '68, 14; medical, 33; microbiome,
33; violent, 124
Rilke, Rainer Maria, 69

Sallis, John, 8
Santayana, George, 9, 57
Sartre, Jean-Paul, 10, 58, 82, 93
Scheffler, Sam, 71
science, 19–20, 24, 34, 39, 94, 108, 131;
and the Anthropocene, 139n2;
climate, 20, 108, 140n8, 150n8; earth,
39; experiments, 91; fiction, 91; hard,
129; Heidegger on, 19, 20, 39, 95,
150n8; hostility to, 109, 129; labora-
tory, 91; laws of, 90, 91; natural, 94;
practice of, 40; reduction to technological control of nature, 40; and
wonder, 37
Scopes "monkey trial," 5
Scotus, Duns, 150n17
Sixth Extinction, 6, 28, 33, 42, 55, 70,
105, 139n2, 143n14
social justice, 123, 133, 134
social media, 128, 129, 134,
social relations, 27, 84, 117, 126, 134
Socrates, 7, 37, 65
sovereignty; human, 28, 29, 30–31, 103;
ecological, 30, 33,

species, 1, 26–27, 35, 38, 49, 83, 104,
139n2; -being, 89; biological, 19;
extinction, 17, 31, 50, 51, 53, 56, 61, 108,
143n13, 152n11; human, 3, 25, 27, 34, 47,
63, 113; -identity, 34; non-human, 3, 28,
33, 51, 71, 72, 83–84; overlap, 11
Spengler, Oswald, 77
Spinoza, Barush, 93, 96, 101,
151n8
Steigler, Bernard, 8
Stoic, 45, 116
structural inertia, 18
subjectivity, 8, 91, 94
the Subject, 21, 29, 41, 68, 100, 106;
agent-, 84, 90; autonomous, 84–85, 88,
89; consumer, 52; position, 21, 102;
species-, 35

techno-hubris, 105
Teilhard de Chardin, Pierre 27, 130–131,
139n4
temporality, 3–8, 10–11, 70
time, 3, 8, 10, 19, 24–25, 70, 126;
aporetic, 19; and boredom, 61;
calculative, 81; cosmic, 60; Derrida
on, 8; geological, 2, 4, 6, 7, 17, 31, 32,
60–61; Heidegger on, 8, 21–22, 74, 76;
linear, 57; lived 6, 10, 15; Nietzsche
on, 8, 19; objectivity of, 3, 4; -shelter,
10, 11; unimaginable, 18
tipping point, 13, 18, 20, 28, 57, 88–89,
132, 134
Toadvine, Ted, 7, 15
Trier, Lars von, 60; *Melancholia*, 18
Tuana, Nancy, 102

uneven development, 18, 27
the Unthinkable, 66, 107, 109, 110–111,
119, 126

Vallega-Neu, Daniela, 8

Watson, James, 34
Wittgenstein, Ludwig, 2, 37, 93
wonder, 16, 36, 37–40, 44–45, 73

Thinking Out Loud: The Sydney Lectures in Philosophy and Society
Dimitris Vardoulakis, series editor

Stathis Gourgouris, *Lessons in Secular Criticism*
Bonnie Honig, *Public Things: Democracy in Disrepair*
David Wood, *Deep Time, Dark Times: On Being Geologically Human*